AF598107

CURRENT TOPICS IN ZIKA

Edited by **Alfonso J. Rodriguez-Morales**

Current Topics in Zika
http://dx.doi.org/10.5772/65145
Edited by Alfonso J. Rodriguez-Morales

Contributors

Pablo Manrique-Saide, Norma Pavia-Ruz, Silvina Contreras-Capetillo, Nina Valadez-González, Ligia Vera-Gamboa, Guillermo Valencia-Pacheco, Rafael Carcaño, Josue Villegas-Chim, Hector Gomez-Dantés, Gonzalo Vazquez-Prokopec, Josue Herrera-Bojorquez, Abdiel Martin-Park, Hugo Delfin-Gonzalez, Fabian Correa-Morales, Dimitrios Vlachakis, Louis Papageorgiou, Vasileios Megalooikonomou, Alfonso J. . Rodriguez-Morales, Syed Badshah, Yahia Mabkhot, Nasir Ahmad, Shazia Syed, Abdul Naeem

Notice

Statements and opinions expressed in the chapters are these of the individual contributors and not necessarily those of the editors or publisher. No responsibility is accepted for the accuracy of information contained in the published chapters. The publisher assumes no responsibility for any damage or injury to persons or property arising out of the use of any materials, instructions, methods or ideas contained in the book.

First published in London, United Kingdom, 2018 by IntechOpen
IntechOpen is the global imprint of INTECHOPEN LIMITED, registered in England and Wales, registration number: 11086078, The Shard, 25th floor, 32 London Bridge Street
London, SE19SG – United Kingdom
Printed in Croatia

British Library Cataloguing-in-Publication Data
A catalogue record for this book is available from the British Library

Additional hard copies can be obtained from orders@intechopen.com

Current Topics in Zika, Edited by Alfonso J. Rodriguez-Morales
p. cm.
Print ISBN 978-1-78923-270-7
Online ISBN 978-1-78923-271-4

For EU product safety concerns:
IN TECH d.o.o., Prolaz Marije Krucifikse Kozulić 3, 51000 Rijeka, Croatia,
info@intechopen.com or visit our website at intechopen.com.

Meet the editor

Dr. Alfonso J. Rodriguez-Morales is an expert in tropical diseases, particularly in zoonotic and vector-borne diseases. His research interests over the recent years have been focused on Zika and Chikungunya, coordinating with the Colombian Collaborative Network on Zika (RECOLZIKA), Chikungunya, and other arboviruses. He is the president of the Travel Medicine Committee of the Pan-American Infectious Diseases Association (API), as well as the secretary of the Colombian Association of Infectious Diseases (ACIN). He is part of the executive board of the Latin American Society for Travel Medicine (SLAMVI). He is a senior researcher at the National Agency of Science in Colombia, Colciencias. Dr. Rodriguez-Morales is a professor and the director of Research, Faculty of Health Sciences, Universidad Tecnológica de Pereira, Pereira, Risaralda, Colombia. His H-index is currently 23.

Contents

Preface IX

Section 1 Clinical Aspects 1

Chapter 1 **Introductory Chapter: Clinical and Epidemiological Implications of Zika Virus Infection - The Experience of RECOLZIKA in Colombia 3**
Alfonso J. Rodríguez-Morales, Adriana M. Trujillo, Jorge A. Sánchez-Duque, Jaime A. Cardona-Ospina, Wilmer E. Villamil-Gómez, Carlos E. Jimenez-Canizalez, Jorge L. Alvarado-Socarras, Juan Carlos Sepúlveda-Arias, Pío López, Matthew Collins, Alberto Paniz-Mondolfi, Antonio C. Bandeira and José Antonio Suárez

Chapter 2 **Zika Virus, Microcephaly and its Possible Global Spread 13**
Syed Lal Badshah, Yahia Nasser Mabkhot, Nasir Ahmad, Shazia Syed and Abdul Naeem

Section 2 Epidemiological and Entomological Aspects 31

Chapter 3 **Genetic and Geo-Epidemiological Analysis of the Zika Virus Pandemic; Learning Lessons from the Recent Ebola Outbreak 33**
Dimitrios Vlachakis, Louis Papageorgiou and Vasileios Megalooikonomou

Chapter 4 **An Integrated Intervention Model for the Prevention of Zika and Other Aedes-Borne Diseases in Women and their Families in Mexico 49**
Norma Pavía-Ruz, Silvina Contreras-Capetillo, Nina Valadéz-González, Josué Villegas-Chim, Rafael Carcaño-Castillo, Guillermo Valencia-Pacheco, Ligia Vera-Gamboa, Fabián Correa-Morales, Josué Herrera-Bojórquez, Hugo Delfín-González, Abdiel Martín-Park, Guadalupe García Escalante, Gonzalo Vázquez-Prokopec, Héctor Gómez-Dantés and Pablo Manrique-Saide

Chapter 5 **Cocirculation and Coinfection Associated to Zika Virus in the Americas 71**
Jorge A. Sánchez-Duque, Alfonso J. Rodríguez-Morales, Adriana M. Trujillo, Jaime A. Cardona-Ospina and Wilmer E. Villamil-Gómez

Preface

Zika was an arbovirus not considered relevant until the epidemics of 2007, where in the islands of the Pacific, Yap, Micronesia, and others, and later in the Americas in 2015–2016, it created a significant public health threat. Zika is a flavivirus that has been especially important not just for the high number of cases but also for its related morbidity. In the case of adult population, multiple neurological diseases are already associated and in pregnant women because of its confirmed teratogenic capacity, leading to microcephaly as well as other central nervous system (CNS) birth defects [1-5]. Many countries in Latin America were significantly affected; in addition to Brazil, this was also the case of Colombia [6-10], where congenital cases (congenital Zika syndrome (CZS)) as well as complicated and even fatal cases were published during the epidemics and beyond [11-15]. However, many aspects are still to be defined, including the impact of co-circulation and coinfection with other arboviruses (e.g., dengue and Chikungunya), where antibody-dependent enhancement (ADE) is also a cause of concern [16-20].

Keeping these issues in mind, this book includes different topics with regard to epidemiology, clinical manifestations, treatment, and prevention of a wide spectrum of manifestations caused by Zika virus in humans, especially including not only the CZS but also entomological aspects. This book has been organized in two major sections: (1) "Clinical Aspects" and (2) "Epidemiological and Entomological Aspects." Section 1 includes topics related to clinical aspects, including particularly the microcephaly and the CZS as well as coinfections. Section 2 includes genetic, geoepidemiological, preventive, and entomological aspects.

Commissioning of this book by IntechOpen team has been related in part to my long commitment with vector-borne, zoonotic, and tropical diseases and being the co-chair of the Working Group on Zoonoses of the International Society for Chemotherapy (WGZ-ISC), as well as in Colombia at the Committee on Zoonoses, Tropical Medicine, and Travel Medicine of the Colombian Association of Infectious Diseases (Asociación Colombiana de Infectología (ACIN)), and more and recently important, the chair of the Colombian Collaborative Network on Zika (Red Colombiana de Colaboración en Zika) (RECOLZIKA), since January 2016.

I have been involved in vector-borne diseases (VBD) for the last two decades, including malaria, leishmaniasis, Chagas disease, as well as dengue and, since 2014, Chikungunya and emerging arboviruses, such as Zika. After moving from Venezuela to Colombia in 2011, I have been involved in the research of VBD in Risaralda, such as malaria and leishmaniasis (still prevalent in the area), where we still keep working on these important tropical diseases. Part of all this is a clear reflection of the work impulse at the Research Group Infection Public Health and Infection (classified as A1 by Colciencias) of the Faculty of Health Sciences of the Universidad Tecnológica de Pereira, directed by Dr. Guillermo Javier Lagos-Grisales, who is not just a partner, a colleague, and a friend but also an extreme believer in our work in vector-borne and zoonotic diseases. But, I must also recognize the beginning of a significant collaboration after a meeting in Cartagena in 2013, during the Colombian Congress of Infectious Diseases, with Dr. Wilmer Ernesto Villamil-Gómez, from Sincelejo, Sucre, Colombia, also part of the former Committee of Zoonoses and Hemorrhagic Fevers of the Colombian Association of Infectious Diseases (Asociación Colombiana de Infectología (ACIN)) (now called the Committee of Zoonoses and Tropical Medicine), who became since that year my most important collaborator on arboviruses, including Zika.

Following the same philosophy as we had on my six previous books with IntechOpen, *Current Topics in Tropical Medicine* [21], *Current Topics in Public Health* [22], *Current Topics in Echinococcosis* [23], *Current Topics in Chikungunya* [24], *Current Topics in Malaria* [25], and *Current Topics in Giardiasis* [26], this book does not intend to be an exhaustive compilation, and this first edition has included not just multiple different topics but also a wide geographical participation from many countries of different regions of the world. Its online availability through the website of IntechOpen, as well as the possibility to upload the complete book or its chapters in personal websites and institution repositories, allows it to reach a wide audience in the globe. Continuing on the series of *Current Topics* books, we are editing *Current Topics in Tropical Emerging Diseases and Travel Medicine*.

I would like to give a very especial thanks to IntechOpen, and particularly to Romina Rovan (Publishing Process Manager), for the opportunity to edit this interesting and important book, as well as for her constant support.

I want to take the appropriate time and space, as I use to do, to dedicate this book to my beloved family (Aurora, Alfonso José, Alejandro, and Andrea (neurologist)) and particularly to Diana, who is more than a wife. After 6 years together, I am so clear she is everything to me. She is the engine of my life. We have gone through difficult moments, partially related to my work, but our love is beyond that. I also would like to acknowledge my friends and my undergraduate and postgraduate students of Health Sciences in Colombia, Venezuela, and around Latin America. I want to thank my colleagues at the Working Group on Zoonoses, International Society for Chemotherapy, and the Committee on Zoonoses, Tropical Medicine, and Travel Medicine (formerly a Committee on Zoonoses and Hemorrhagic Fevers) of the Colombian Association of Infectious Diseases (ACIN) and a large list of the members of RECOLZIKA (http://blog.utp.edu.co/arodriguezm/zika/). Special thanks to my friend and colleague Dr. Guillermo J. Lagos-Grisales, MD, MPH, and to the members of our research group and incubator consisting of young and enthusiastic medical students and some veterinary medical students as well as young medical doctors, who are pursuing significant improvements in the understanding of the epidemiology of zoonotic, vector-borne, parasitic, and in general infectious diseases in our country with international projection. Year 2017 has been highly productive for this recognized group, which now was classified by the National Agency of Science, Colciencias, in the highest rank "A1," which is positioning as a leader in infectious disease epidemiology research in the Coffee Triangle region and in the country.

Finally, I hope our readers enjoy this publication as much as I did reading the chapters of *Current Topics in Zika*.

Prof. Alfonso J. Rodriguez-Morales
MD, MSc, DTM&H, FRSTMH(Lon), FFTM RCPS(Glasg), FACE, PhD(c)
Editor, Current Topics in Zika
Part-Time Faculty Instructor, Risk Factors (Epidemiology) (Coordinator)
Department of Community Medicine, School of Medicine;
Lecturer, Frontiers Research, School of Veterinary Medicine & Zootechnics
Director of Research, Faculty of Health Sciences, Universidad Tecnológica de Pereira
Pereira, Risaralda, Colombia
Chair, Colombian Network of Research in Zika (RECOLZIKA)
Co-Chair, Working Group on Zoonoses, International Society for Chemotherapy (ISC)
Secretary, Colombian Association of Infectious Diseases (ACIN)
Member of the Committee on Zoonoses, Tropical Medicine, and Travel Medicine, ACIN
President, ACIN Chapter Coffee Triangle Region
President, Committee on Travel Medicine
Pan-American Association of Infectious Diseases
Senior Researcher, Colciencias (2017-2019)

References

[1] Rodríguez-Morales AJ. Zika: the new arbovirus threat for Latin America. J Infect Dev Ctries 2015 Jun; 9(6):684-685.

[2] Martinez-Pulgarin DF, Acevedo-Mendoza WF, Cardona-Ospina JA, Rodríguez-Morales AJ, Paniz-Mondolfi AE. A bibliometric analysis of global Zika research. Travel Medicine & Infectious Disease 2016 Jan-Feb; 14(1):55-57.

[3] Villamil-Gómez WE, González-Camargo O, Rodriguez-Ayubi J, Zapata-Serpa D, Rodríguez-Morales AJ. Dengue, Chikungunya and Zika co-infection in a patient from Colombia. J Infect Public Health 2016 2016 Sep-Oct;9(5):684-686.

[4] Arzuza-Ortega L, Polo A, Pérez-Tatis G, López-García H, Parra E, Pardo-Herrera LC, Rico-Turca AM, Villamil-Gómez W, Rodríguez-Morales AJ. Fatal Sickle Cell Disease and Zika Virus Infection in Girl from Colombia. Emerg Infect Dis 2016 May; 22(5):925-927.

[5] Rodríguez-Morales AJ. Zika and Microcephaly in Latin America: an emerging threat for pregnant travelers? [Editorial] Travel Medicine & Infectious Disease 2016 Jan-Feb; 14(1):5-6.

[6] Rodriguez-Morales AJ, Villamil-Gómez W. El Reto de Zika en Colombia y América Latina: Una emergencia sanitaria internacional. Infectio 2016 Abr-Jun; 20(2): 59-61.

[7] Rodriguez-Morales AJ, Bandeira AC, Franco-Paredes C. The expanding spectrum of modes of transmission of Zika virus: a global concern. Ann Clin Microbiol Antimicrob. 2016 Mar 3; 15:13.

[8] Villamil-Gómez WE, Mendoza-Guete A, Villalobos E, González-Arismendy E, Uribe-García AM, Castellanos JE, Rodriguez-Morales AJ. Diagnosis, Management and Follow-up of Pregnant Women with Zika virus infection: A preliminary report of the ZIKERNCOL cohort study on Sincelejo, Colombia. Travel Medicine & Infectious Disease 2016 Mar-Apr; 14(2):155-158.

[9] Sarmiento-Ospina A, Vásquez-Serna H, Jimenez-Canizales CE, Villamil-Gómez WE, Rodriguez-Morales AJ. Zika virus associated deaths in Colombia. Lancet Infect Dis 2016 May; 16(5):523-524.
[10] Cardona-Cardona AF, Rodríguez-Morales AJ. Severe abdominal pain in a patient with Zika infection: a case in Risaralda, Colombia. J Infect Public Health 2016 May–Jun; 9(3):372–373.
[11] Rodríguez-Morales AJ, Villamil-Gómez WE, Franco-Paredes C. The arboviral burden of disease caused by co-circulation and co-infection of dengue, chikungunya and Zika in the Americas. Travel Medicine & Infectious Disease 2016 May-Jun; 14(3):177-179.
[12] Villamil-Gómez WE, Rodríguez-Morales AJ, Uribe-García AM, González-Arismendy E, Castellanos JE, Calvo EP, Álvarez-Mon M, Musso D. Zika, Dengue and Chikungunya Co-Infection in a Pregnant Woman from Colombia. Int J Infect Dis 2016 Oct; 51(10):135-138.
[13] Rodriguez-Morales AJ, Galindo-Marquez ML, García-Loaiza CJ, Sabogal-Roman JA, Marin-Loaiza S, Ayala AF, Lagos-Grisales GJ, Lozada-Riascos CO, Parra-Valencia E, Rojas-Palacios JH, López E, López P, Grobusch MP. Mapping Zika virus disease incidence in Valle del Cauca. Infection 2017 Feb; 45(1):93-102.
[14] Rodríguez-Morales AJ, Anaya J-M. Impacto de las arbovirosis artritogénicas emergentes en Colombia y América Latina. Revista Colombiana de Reumatología 2016 Jul-Sep; 23(3):145-147.
[15] Villamil-Gomez WE, Sánchez-Herrera AR, Hernandez H, Hernández-Iriarte J, Díaz-Ricardo K, Castellanos J, Villamil-Macareno WJ, Rodriguez-Morales AJ. Guillain-Barré syndrome during the Zika virus outbreak in Sucre, Colombia, 2016. Travel Med Infect Dis 2017 Mar-Apr; 16(2):62-63.
[16] Paniz-Mondolfi AE, Hernandez-Perez M, Blohm G, Marquez M, Mogollon-Mendoza A, Hernandez-Pereira CE, DeArocha JR, Rodriguez-Morales AJ. Generalized pustular psoriasis triggered by Zika virus infection. Clinical and Experimental Dermatology 2017 Epub Ahead Oct 13; http://onlinelibrary.wiley.com/doi/10.1111/ced.13294/full
[17] Blohm G, Lednicky J, Marquez M, White S, Loeb J, Pacheco C, Nolan D, Paisie T, Salemi M, Rodriguez-Morales AJ, Morris JG, Pulliam J, Carrillo A, Plaza J, Paniz-Mondolfi A. Complete Genome Sequences of Identical Zika Viruses in a Nursing Mother and Her Infant. Genome Announc. 2017 Apr; 5(17):e00231-17.
[18] Rodriguez-Morales AJ, Ruiz P, Tabares J, Ossa CA, Yepes-Echeverry MC, Ramirez-Jaramillo V, Galindo-Marquez ML, García-Loaiza CJ, Sabogal-Roman JA, Parra-Valencia E, Lagos-Grisales GJ, Lozada-Riascos CO, de Pijper CA, Grobusch MP. Mapping the Ecoepidemiology of Zika virus infection in Urban and Rural Areas of Pereira, Risaralda, Colombia, 2015-2016: Implications for Public Health and Travel Medicine. Travel Med Infect Dis 2017 Jul-Aug; 18(C):57-66.
[19] Villamil-Gómez WE, Guijarro E, Castellanos J, Rodriguez-Morales AJ. Congenital Zika Syndrome with Prolonged Detection of Zika Virus RNA. J Clin Virol 2017 Oct; 95(10):52-54.

[20] Karam E, Giraldo J, Rodriguez FM, Hernández-Pereira CE, Rodriguez-Morales AJ, Blohm G, Paniz-Mondolfi A. Ocular flutter following Zika virus infection. Journal of NeuroVirology 2017 Epub Ahead Nov 16; https://link.springer.com/article/10.1007%2Fs13365-017-0585-1

[21] Rodriguez-Morales AJ. (Editor). Current Topics in Tropical Medicine. ISBN 978-953-51-0274-8. InTech, Croatia, March 2012. Available at: http://www.intechopen.com/books/current-topics-in-tropical-medicine

[22] Rodriguez-Morales AJ. (Editor). Current Topics in Public Health. ISBN 978-953-51-1121-4. InTech, Croatia, May 2013. Available at: http://www.intechopen.com/books/current-topics-in-public-health

[23] Rodriguez-Morales AJ. (Editor). Current Topics in Echinococcosis. ISBN 978-953-51-2159-6. InTech, Croatia, September 2015. http://www.intechopen.com/books/current-topics-in-echinococcosis

[24] Rodriguez-Morales AJ. (Editor). Current Topics in Chikungunya. ISBN 978-953-51-2595-2. InTech, Croatia, August 2016. http://www.intechopen.com/books/current-topics-in-chikungunya

[25] Rodriguez-Morales AJ. (Editor). Current Topics in Malaria. ISBN 978-953-51-2790-1. InTech, Croatia, November 2016. http://www.intechopen.com/books/current-topics-in-malaria

[26] Rodriguez-Morales AJ. (Editor). Current Topics in Malaria. ISBN 978-953-51-5260-6. InTech, Croatia, December 2017. http://www.intechopen.com/books/current-topics-in-giardiasis

Section 1

Clinical Aspects

Chapter 1

Introductory Chapter: Clinical and Epidemiological Implications of Zika Virus Infection - The Experience of RECOLZIKA in Colombia

Alfonso J. Rodríguez-Morales, Adriana M. Trujillo, Jorge A. Sánchez-Duque, Jaime A. Cardona-Ospina, Wilmer E. Villamil-Gómez, Carlos E. Jimenez-Canizalez, Jorge L. Alvarado-Socarras, Juan Carlos Sepúlveda-Arias, Pío López, Matthew Collins, Alberto Paniz-Mondolfi, Antonio C. Bandeira and José Antonio Suárez

Additional information is available at the end of the chapter

http://dx.doi.org/10.5772/intechopen.72571

1. Introduction

1.1. Overview and general aspects of Zika

Zika virus (ZIKV) is a mosquito-borne flavivirus discovered in rhesus monkeys in the Zika close to Kampala, Uganda in 1947, but it was not until February 2016 that the World Health Organization (WHO) declared Zika a public health emergency [1–5]. The introduction and spread of ZIKV throughout Latin America embodies a convergence of ecologic, social and environmental factors that foster the emergence of new infectious threats in susceptible populations. Although ZIKV infection is typically asymptomatic or causes a mild flu-like, birth defects indicate a wide clinical spectrum that includes severe manifestations that underlie global concern [4–7]. These, along with evidence of non-vector-borne transmission routes such as vertical transmission, blood transfusion and sexual contact [8–11], highlight the need for augmented research in the area [12]. Currently, research efforts have been increasingly focusing on disease prevention through vaccine design and understanding of the role of antibody disease enhancement (ADE) on the immunopathogenesis of severe cases and fetal outcomes. New diagnostic tools are required in order to bypass cross reactivity with other flaviviruses

like dengue or yellow fever, and there is a gap for implementation research in order to develop effective strategies for vector control and awareness programs among people [10–18].

2. Epidemiology

It is difficult to exaggerate the medical importance and burden of vector-borne infectious diseases as a series of emerging and re-emerging arboviruses epidemics are propagated in recent decades in previously unexposed geographic areas in Latin America and the Caribbean [9]. ZIKV has caused epidemics during 2015–2017 in different countries of the Americas, with more than 50 countries/territories affected in this region, and 148 globally affected in any form (including imported cases). After Brazil, probably Colombia is the second most affected country in the region, with over 100,000 cases having been reported from this northern South American country, reflecting overall incidence rates above 150 cases/100,000 [18]. During the years 2015–2017, the circulation of ZIKV was confirmed in 560 municipalities and four districts of Colombia. Suspected cases of ZIKV disease have been reported in 245 municipalities, adding a total of 809 municipalities with reported cases between confirmed and suspected cases. Thirty-five territorial entities of the departmental and district order have notified the surveillance system of ZIKV [19].

Although Colombia is not currently (November 2017) at epidemic status for ZIKV, it has become endemic with continued transmission. The main territories affected are those with previous circulation of dengue and chikungunya. In this sense, the departments (at the first administrative level) of Valle del Cauca, Santander, Tolima, Cundinamarca and Meta account with more than the 60% of the cases reported during 2017. On the other hand, 238 cases of symptomatic ZIKV infection had been documented in pregnant women (37 confirmed cases), and 20 cases charted in municipalities with no previously known ZIKV transmission. By territorial entity of residence, the one that has reported the largest proportion of cases is Santander with 46 pregnant women in the northeastern region, border with Venezuela [19].

3. Clinical aspects

It is estimated that 80% of people infected by ZIKV are asymptomatic, but they can still develop complications that lead the patient to death or generate severe chronic conditions such as Guillain-Barré syndrome. Asymptomatic individuals may be important for sustaining ZIKV transmission within a population despite being undiagnosed as a ZIKV case [6, 20].

Individuals with acute, symptomatic cases (25%) experience fever (elevation of axillary body temperature greater than 37.2°C), nonpurulent conjunctivitis, headache, myalgia, arthralgia, asthenia, maculopapular rash (usually in extremities and trunk) **(Figure 1)**, lower limb edema and less frequently, retro-orbital pain, and gastrointestinal disturbance such as abdominal pain, nausea, and diarrhea. Notwithstanding, afebrile patients with rash should also be assessed for ZIKV infection [3, 11, 12]. These relatively mild symptoms last a few days (4–7 days), uncommonly result in hospitalization and are hardly distinguishable from other, better-known disease caused by other arboviruses [7, 16, 21]. Some neurological manifestations including Guillain-Barré syndrome,

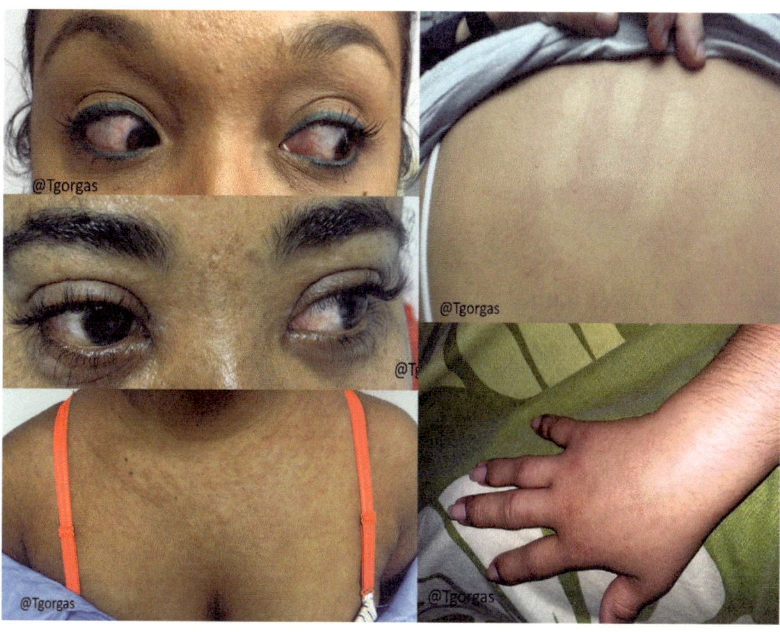

Figure 1. Clinical aspects of patients with Zika virus infection (confirmed by PCR) (pictures took and provided with patient authorization, by co-author Jose Antonio Suarez, Instituto Gorgas, Panama): conjunctivitis, rash, arthralgia.

acute myelitis, and meningoencephalitis are associated with detection of ZIKV in the cerebrospinal fluid [2, 16, 22]. Interestingly, this virus can cross the fetoplacental barrier to affect the fetus and cause microcephaly, microcephaly-related intellectual disabilities to neonates, ophthalmological alterations, loss hearing and epilepsy. Evidence has suggested that ZIKV could also invade cardiac cells, which could explain a described association between congenital heart disease in an infant born to a ZIKV-infected mother [13, 23, 24]. Differentiation on clinical grounds alone is often a very difficult task. The biological and clinical behaviors exhibit not only different characteristics but also great similarities; thus, ZIKV could be easily confused with other arboviruses. The clinical picture may be further complicated by coinfections with various combinations of arboviruses [9, 12, 18].

4. Diagnostics

The transmission of the diseases through mosquito bite is the most common scenario in ZIKV infection, which was described in 1966 in Malaysia. However, few cases of nonvector-borne infections have been reported, and the existence of ZIKV in the pharynx and saliva of infected patients represents a potential but unproven route of transmission. ZIKV also has been detected in semen, cervicovaginal fluid and urine, consistent with several well-documented cases of sexual transmission [3, 14, 16, 25].

The diagnosis of ZIKV infection is established through detection of ZIKV RNA by reverse transcription polymerase chain reaction (RT-PCR) typically performed on serum, plasma or urine within 2 weeks of symptom onset. ZIKV can also be diagnosed by serology (IgM antibodies). Serum RT-PCR is positive if done in the acute phase of the viremia, in blood on the first 3–7 days of the onset of illness, and for up to 10–14 days in urine. Enzyme-linked immunosorbent assay (ELISA) is used to detect IgM, but unfortunately, there are only a few laboratories able to perform an ELISA for ZIKV. Fewer laboratories still are able to perform neutralization assays, which may be more specific than ELISA, but require greater resources and biosafety containment. Although the specific antibodies against ZIKV in serum (IgM antibodies) are detectable after 4 days of symptom onset, its diagnostic value is limited due to cross-reactivity with other flaviviruses. Comparing relative titers of binding or neutralizing antibodies may be helpful in discriminating ZIKV from related flaviviruses like dengue in research settings, this is not yet standardized to aid in clinical diagnosis [16, 21].

The probable modes of perinatal transmission are transplacental or may occur during delivery. Concern has raised about the possibility of breastfeeding transmission. Prenatal fetal evaluation of pregnant women suspected or confirmed to have a ZIKV infection is done by regular fetal ultrasound examination, which can detect abnormalities as early as 18–20 weeks of gestation. The main ultrasound findings associated with fetal ZIKV infection are microcephaly, intracranial calcifications, hydranencephaly, ventricular dilatation, brain atrophy, anhydramnios, *hydrops fetalis*, and intrauterine growth retardation. Amniotic fluid obtained by amniocentesis after 15 weeks of gestation can be tested for the presence of viral RNA by RT-PCR, but the sensitivity and specificity is unknown during gestation [13, 21, 26].

It is important in the differential diagnosis of dengue and chikungunya, among other conditions such as malaria, leptospirosis, measles, and also to consider the possibility of coinfections, especially in endemic areas where all these pathogens can cocirculate simultaneously. During the symptomatic period of infection, various laboratory parameters provide information on ZIKV infection. Various laboratory parameters such as leucopenia, thrombocytopenia, serum lactate dehydrogenase, gamma glutamyl transferase and elevated protein markers may be suggestive of ZIKV infection, but these findings are nonspecific [4, 13, 26]. Also, the virus can be detected in semen up to 81 days after infection.

5. Treatment

Currently, there is no ZIKV vaccine available, but the WHO has made ZIKV vaccine development a top priority. Thus, more than 50 ZIKV vaccine candidates are now in various stages of research and development, mainly, in phase I/II clinical trials that include inactivated whole viruses, recombinant measles viral vector-based vaccines, DNA and mRNA vaccines, and a mosquito salivary peptide vaccine [17, 27].

The treatment for ZIKV infection is entirely supportive, no antiviral. No drugs have yet been approved for a specific ZIKV treatment, although numerous nucleoside analogs have some antiviral activities in cell cultures such as ribavirin, sofosbuvir or favipiravir. Acetaminophen is used to control fever and pain, avoiding aspirin or nonsteroidal anti-inflammatory drugs, because of their risk of hemorrhage (in case of DENV infection), and fluids are generously administered to prevent dehydration [24, 26, 28].

6. Control and prevention

During the last few decades, Latin America has been threatened by an unprecedented explosion of emerging arboviral outbreaks. These epidemics of emerging and re-emerging arboviruses are due to a number of factors such as climate change, levels of urbanization, increasing international travel, foreign trade, poor socioeconomic conditions, susceptible geographical areas (tropical and subtropical regions) among other factors. The presence of *Aedes* mosquitos enables ZIKV to invade new areas and poses a worldwide risk as there are no preventive approaches or vaccines for ZIKV [9, 12, 16, 26].

In countries where ZIKV epidemics are reported, cost-effectiveness studies may be important to determine the feasibility of systematic screening of blood components in donated samples. Unfortunately, at the moment, no specific data are available on the rate of reduction of blood-borne transmission due to such practices. Because the prevalence of viremia in blood donations is high in endemic areas, mainly asymptomatic donors, donor selection professionals should carefully evaluate the appropriateness of the donation based on the patient's background, likewise, follow up on donors with the aim of identifying the onset of symptoms [16, 26].

The most challenging aspect of ZIKV is preventing congenital infection. Recommendations to avoid mosquito exposure or avoid pregnancy altogether are not practical for most women living in ZIKV-endemic areas. Women who are or are planning to become pregnant are discouraged from visiting areas of ZIKV transmission by travel restrictions issued by the US Center for Disease Control and other public health institutions. For preventing the sexual transmission of ZIKV for couples in which a man has traveled to or resides in an area with active ZIKV transmission, safe sexual practices for 6 months after the exposure regardless of the appearance of symptoms are recommended. What constitutes totally safe sexual practice is unclear, and some studies suggest that transmission of the small ZIKV (40 nm) may not be fully prevented by condoms [1, 2, 26, 29].

Much work remains before effective antiviral drugs and vaccines are available for ZIKV. For the time being, recognizing the importance of epidemiological control of emerging viral diseases and taking preventive measures such as the increase in *Aedes* vector control and containment strategies guided by the scientific literature is a priority to achieve control of tropical viruses with epidemic potential [2, 3, 9]. In general, in countries with sporadic imported cases of ZIKV infection or in ZIKV-free countries, the only precaution that must be taken is to notify the cases. However, for epidemic areas, the main way to combat ZIKV is to cover skin by cloths, use mosquito repellents, reside in air-conditioned or screened rooms, and avoid being near water containers like stagnant water ponds, old automobile tire or plants containing water [16, 17, 26].

Finally, it is important to note that because only symptomatic individuals are diagnosed with ZIKV, robust epidemiologic data required for optimal control programs are lacking. Additionally, only a few reported cases have received laboratory confirmation; thus, conditions such as dengue, chikungunya, malaria, leptospirosis, measles, among others may be incorrectly identified as ZIKV (or vice versa).

Author details

Alfonso J. Rodríguez-Morales[1,2,3,4,5,6*], Adriana M. Trujillo[1], Jorge A. Sánchez-Duque[1], Jaime A. Cardona-Ospina[1,2,6], Wilmer E. Villamil-Gómez[3,6,7,8], Carlos E. Jimenez-Canizalez[1,6,9], Jorge L. Alvarado-Socarras[6,10], Juan Carlos Sepúlveda-Arias[2,6], Pío López[6,11,12], Matthew Collins[13], Alberto Paniz-Mondolfi[6,14,15], Antonio C. Bandeira[6,16] and José Antonio Suárez[6,17]

*Address all correspondence to: arodriguezm@utp.edu.co

1 Public Health and Infection Research Group, Faculty of Health Sciences, Universidad Tecnológica de Pereira, Pereira, Risaralda, Colombia

2 Infection and Immunity Research Group, Faculty of Health Sciences, Universidad Tecnológica de Pereira, Pereira, Risaralda, Colombia

3 Committee on Zoonoses, Tropical Medicine and Travel Medicine, Asociación Colombiana de Infectología, Bogotá, DC, Colombia

4 Working Group on Zoonoses, International Society for Chemotherapy, Aberdeen, United Kingdom

5 Committee on Travel Medicine, Pan-American Association of Infectious Diseases, Panama City, Panama

6 Colombian Collaborative Network for Research on Zika and other Arboviruses (RECOLZIKA), Pereira, Risaralda, Colombia

7 Infectious Diseases and Infection Control Research Group, Hospital Universitario de Sincelejo, Sincelejo, Sucre, Colombia

8 Programa del Doctorado de Medicina Tropical, SUE Caribe, Universidad del Atlántico, Barranquilla, Colombia

9 Department of Internal Medicine, Universidad Surcolombiana, Neiva, Huila, Colombia

10 Unidad de Neonatología, Departamento de Pediatría, Fundación Cardiovascular de Colombia, Bucaramanga, Santander, Colombia

11 Departamento de Pediatría, Hospital Universitario del Valle, Universidad del Valle, Cali, Valle del Cauca, Colombia

12 Centro de Estudios en Infectología Pediátrica, Cali, Valle del Cauca, Colombia

13 Department of Medicine, Division of Infectious Diseases, Emory University, Atlanta, GA, USA

14 Department of Tropical Medicine and Infectious Diseases, Hospital Internacional Barquisimeto, Lara, Barquisimeto, Venezuela

15 Instituto Venezolano de los Seguros Sociales (IVSS), Department of Health, Caracas, Venezuela

16 Hospital Aliança and Faculdade de Tecnologia e Ciencias Medical School, Salvador, Brazil

17 Clinical Research Department, Instituto Conmemorativo Gorgas de Estudios de la Salud, Panama City, Panama

References

[1] Rodriguez-Morales AJ, Espinoza-Flores LA. Should we be worried about sexual transmission of Zika and other arboviruses? International Maritime Health. 2017;**68**(1):68-69

[2] Rodríguez-Morales AJ, Willamil-Gómez WE. El reto de Zika en Colombia y América Latina: Una urgencia sanitaria internacional. Infection. 2016;**20**(2):59-61

[3] Rodriguez-Morales AJ. Zika: The new arbovirus threat for Latin America. The Journal of Infection in Developing Countries. 2015;**9**(06):684-685

[4] Rodriguez-Morales AJ, Acevedo WF, Villamil-Gómez WE, Escalera-Antezana JP. Aspectos Clínicos y Epidemiológicos de la Infección por Virus Zika: Implicaciones de la Actual Epidemia en Colombia y América Latina. Hechos Microbiológicos. 2016;**5**(2):92-105

[5] Gatherer D, Kohl A. Zika virus: A previously slow pandemic spreads rapidly through the Americas. Journal of General Virology. 2016;**97**(2):269-273

[6] Patiño-Barbosa AM, Rodríguez-Morales AJ. Debemos esperar una mayor expansión de distintos Arbovirus en las Américas. Ciencia e Investigación Medico Estudiantil Latinoamericana. 2017;**22**(2):2-8

[7] Rodriguez-Morales AJ. No era suficiente con dengue y chikungunya: llegó también Zika dengue and chikungunya were not enough: Now also Zika. Archivos de Medicina. 2015;**11**(2):3

[8] Rodríguez-Morales AJ, Sánchez-Duque JA, Anaya J-M. Respuesta a: Alfavirus tropicales artritogénicos. Reumatología Clínica. 2017. DOI: 10.1016/j.reuma.2017.07.011

[9] Rodríguez-Morales AJ, Sánchez-Duque JA. Preparing for next arboviral epidemics in Latin America; who can it be now? – Mayaro, Oropouche, West Nile or Venezuelan Equine Encephalitis viruses. Journal of Preventive Epidemiology. 2017;**3**(1):e01

[10] Rodríguez-Morales AJ, Anaya J-M. Impacto de las arbovirosis artritogénicas emergentes en Colombia y América Latina. Revista Colombiana de Reumatologia. 2016;**23**(3):145-147

[11] Villamil-Gómez WE, González-Camargo O, Rodriguez-Ayubi J, Zapata-Serpa D, Rodriguez-Morales AJ. Dengue, chikungunya and Zika co-infection in a patient from Colombia. Journal of infection and public health. 2016;**9**(5):684-686

[12] Paniz-Mondolfi AE, Rodriguez-Morales AJ, Blohm G, Marquez M, Villamil-Gomez WE. ChikDenMaZika syndrome: The challenge of diagnosing arboviral infections in the midst of concurrent epidemics. Annals of Clinical Microbiology and Antimicrobials. 2016; **15**(1):42

[13] Waggoner JJ, Pinsky BA. Zika virus: Diagnostics for an emerging pandemic threat. Journal of Clinical Microbiology. 2016;**54**(4):860-867

[14] Rodriguez-Morales AJ, Bandeira AC, Franco-Paredes C. The expanding spectrum of modes of transmission of Zika virus: A global concern. Annals of Clinical Microbiology and Antimicrobials. 2016;**15**(1):13

[15] Rodríguez-Morales A. *Aedes*: un eficiente vector de viejos y nuevos arbovirus (dengue, chikungunya y zika) en las Américas. Revista del Cuerpo Médico del HNAAA. 2015; **8**(2):50-52

[16] Marano G, Pupella S, Vaglio S, Liumbruno GM, Grazzini G. Zika virus and the never-ending story of emerging pathogens and transfusion medicine. Blood Transfusion. 2016; **14**(2):95

[17] Wahid B, Ali A, Rafique S, Idrees M. Current status of therapeutic and vaccine approaches against Zika virus. European Journal of Internal Medicine. 2017;**44**(10):12-18

[18] Rodriguez-Morales AJ, Ruiz P, Tabares J, Ossa CA, Yepes-Echeverry MC, Ramirez-Jaramillo V, et al. Mapping the ecoepidemiology of Zika virus infection in urban and rural areas of Pereira, Risaralda, Colombia, 2015-2016: Implications for public health and travel medicine. Travel Medicine and Infectious Disease. 2017;Jul - Aug;**18**:57-66

[19] INS. Boletín Epidemiológico Semanal. In: Pública DdVyAdReS, editor. Instituto Nacional de Salud. 2017. Available from: www.ins.gov.co

[20] Rodriguez-Morales AJ. Zika and microcephaly in Latin America: An emerging threat for pregnant travelers? Travel Medicine and Infectious Disease. 2016;**14**(1):5

[21] Plourde AR, Bloch EM. A literature review of Zika virus. Emerging Infectious Diseases. 2016;**22**(7):1185

[22] Carteaux G, Maquart M, Bedet A, Contou D, Brugières P, Fourati S, et al. Zika virus associated with meningoencephalitis. New England Journal of Medicine. 2016;**374**(16): 1595-1596

[23] Minhas AM, Nayab A, Shobana Iyer MN, Fatima K, Khan MS, Constantin J. Association of Zika virus with myocarditis, heart failure, and arrhythmias: A literature review. Cureus. 2017;**9**(6):e1399

[24] Radha B, Muniraj G. Alternate paradigms on Zika virus-related complications: An analytical review. Asian Pacific Journal of Tropical Medicine. 2017;**10**(7):631-634

[25] Martinez-Pulgarin DF, Acevedo-Mendoza WF, Cardona-Ospina JA, Rodríguez-Morales AJ, Paniz-Mondolfi AEA. Bibliometric analysis of global Zika research. Travel Medicine and Infectious Disease. 2016;**14**(1):55-57

[26] Shankar A, Patil AA, Skariyachan S. Recent perspectives on genome, transmission, clinical manifestation, diagnosis, therapeutic strategies, vaccine developments, and challenges of Zika virus research. Frontiers in Microbiology. 2017;**8**:1761

[27] Lin H-H, Yip B-S, Huang L-M, Wu S-C. Zika virus structural biology and progress in vaccine development. Biotechnology Advances. 2017. DOI: 10.1016/j.biotechadv.2017.09.004

[28] Munjal A, Khandia R, Dhama K, Sachan S, Karthik K, Tiwari R, et al. Advances in developing therapies to combat Zika virus: Current knowledge and future perspectives. Frontiers in Microbiology. 2017;**8**:1469

[29] Meaney-Delman D, Rasmussen SA, Staples JE, Oduyebo T, Ellington SR, Petersen EE, et al. Zika virus and pregnancy: What obstetric health care providers need to know. Obstetrics & Gynecology. 2016;**127**(4):642-648

Chapter 2

Zika Virus, Microcephaly and its Possible Global Spread

Syed Lal Badshah, Yahia Nasser Mabkhot,
Nasir Ahmad, Shazia Syed and Abdul Naeem

Additional information is available at the end of the chapter

http://dx.doi.org/10.5772/intechopen.72507

Abstract

Zika virus is an arbovirus that is spreading at an alarming state in the American continents and now in Asian countries. The Aedes mosquitoes are the vectors for the spread of this virus beside other ways of transmission. Currently, there are no vaccines or drugs available for its treatment. The Zika virus–related microcephaly cases are reported in fetuses of pregnant women who got this viral infection. However, the exact mechanism of Zika virus and microcephaly is still not established. Here we review Zika virus epidemiology, its unusual relationship with microcephaly in fetuses and current scientific research progress on it.

Keywords: Zika virus, microcephaly, Aedes mosquitos, epidemic, vaccines

1. Introduction and background

Zika virus (ZIKV) of the family Flaviviridae was on the list of neglected diseases, and like all neglected diseases, it is now an epidemic causing microcephaly in new-born and Guillain-Barré syndrome in adults [1]. The *Aedes aegypti* mosquitoes are the main vectors for it beside other reported species like *Aedes hensilli* [2]. Historically, it was identified back in 1947, and recently, Chang et al. have thoroughly discussed its full history [3] and several authors have reviewed the literature [4–9], but still there is a need for updated information. In the past 70 years, different strains of Zika virus have been isolated from time to time in different parts of Africa [10–15]. It is generally believed that the current progress in human travel technology and its frequent use by people and the spread of Aedes due to its invasive nature cause the global spread of Zika virus [6]. Zika virus infection was also reported in a few returning travelers in the Netherland and Switzerland, so its possible spread to the Europe is also difficult to avoid [16–18]. It was recently reported that another vector mosquito called *Aedes albopictus* has the capacity to spread Zika virus in North America and Europe [19].

The threat of Zika virus to the United States population was not given importance initially, but several reports from Puerto Rico, Texas, and other parts alarm bells in the scientific community and a wake-up call to National Institute of Health [20, 21]. Zika virus is now on the door step of Asia with cases reported from Manila (Philippines), and parts of Singapore, Thailand, India, etc., which have favorable environmental conditions for Aedes mosquitos, and in the past, minor outbreaks were also recorded [14, 22–24]. This is an alarming situation and there is no preparation on the emergency basis. In the past, several cases of Zika virus were reported in Asian countries like Cambodia [25] and Senegal in Africa [26, 27], but it was not that widespread. Before 2007, few cases were reported at different places in different times, but it is suggested that increase in mosquito breeding and environmental changes due to human activities may cause mutation in the RNA genome of Zika virus and it becomes more adaptable to the mosquitos [28, 29]. In 2014–2015, Zika virus infection spread in South and Central America and Caribbean regions and is now spreading to other parts of the world [20, 30]. The spread of Zika virus in northeastern parts of Brazil occurred in the beginning of 2015 when it was observed that some patients had mild fever, rash, conjunctivitis and arthralgia and their tests were negative for dengue virus [31]. Later on, the reverse transcription-polymerase chain reaction from the sera showed the presence of Zika virus and its classification showed that it belonged to the Asian group [31]. The human skin fibroblasts cells are permissive to Zika virus entry. The viral replication in infected cells causes activation of an antiviral innate immune response with the production of type I interferons [32].

2. Transmission

The common possible ways of spreading beside the mosquito bites and transplacental transmission are blood transfusion, transplantation of organs, sexual activities, breastfeeding [33], respiratory along with sweat and tear droplets and animal bite [28, 34]. In North America and European countries, the spread through sexual contacts are more as these regions are not favorable for Aedes mosquitos [35]. In such areas, blood and semen donations should be properly checked for the presence of Zika virus, and further, people coming from affected areas should take care when they donate blood and semen in unaffected areas [36].

3. Relation between Zika virus and microcephaly

Microcephaly is one of the associated medical effects of Zika virus infection in mothers with their new born babies [36–40]. In the past, cases of Zika virus with microcephaly were not that common, but in Brazil, it was observed that the rate of microcephaly is twenty times more, which is quite alarming and new cases are still emerging [41]. Zika virus has the capacity to cross the placental barrier and infect the fetus [42, 43]. The infected mothers of these babies have rashes in the first and second trimesters [44]. Beside microcephaly, brain calcifications, cataracts and intraocular calcifications of eyes in fetuses are also reported [45]. In other cases, agyria, hydrocephalus and associated cortical displacement and mild focal inflammation are also present [46–48]. Similarly in another case in addition to the above-mentioned symptoms, ultrasound of the infected fetus showed damaged lesions of posterior fossa and ascites along with subcutaneous edema [49]. Recent studies also suggested that the brain tissue and cells are favorable places for Zika virus growth as it targets the brain tissues in the neonate very easily [50, 51]. Zika virus decreases the viability and growth of neurospheres and brain organoids, and in that way, it targets the brain to cause microcephaly [52, 53]. Bullerdiek et al. hypothesized that the virus proteins affect the mitotic spindle proteins and their apparatus and this could be the possible way through which Zika virus causes teratogenic effects like microcephaly [54].

There are also ambiguities about the role of Zika virus and microcephaly, as in the past microcephaly was not linked with Zika virus infection, and the cases of microcephaly were very low and it is possible that other factors may be involved in it [55, 56]. Evans et al. suggested that the insecticide pyriproxyfen that is used for the control of mosquito populations in drinking water is the possible agent for microcephaly in Brazil [57]. The insecticide pyriproxyfen has shown to act as juvenile hormone and has cross-reactivity with most of the fat soluble molecules like retinoic acid (a metabolite of vitamin A) and a ligand for RXR receptor [57]. The RXR receptor is a type of nuclear receptor and this cross-reactivity of two agents results into abnormal activity of RXR receptors with the ultimate abnormality of microcephaly [57]. However, experiments on zebra fish models showed that pyriproxyfen did not cause any deformities in the central nervous system [58]. To counter such ambiguities, different groups are studying the effect of Zika virus on mouse models, and in one of the recent reports, the researchers showed that when a mouse is infected with a Brazilian Zika virus strain, it causes intrauterine growth restriction and signs of microcephaly were also observed [59]. They also showed that Zika virus attacks the human cortical progenitor cells and destroys them, and further the viral infection also damages the human brain organoids with a decrease in proliferative zones and damages the cortical layers [59]. In different brain and neural cells infected with Zika virus, it has been observed that the viruses halt the process of mitoses and damage the centrosome and the structural organization of the dividing cell with ultimate result of cell death [60]. It has been observed that a single point mutation of serine to asparagine (S139 N) in the ZIKV polyprotein resulted in high virulence ability in human progenitor cells [61]. This mutation occurred before the 2013 French Polynesia and was maintained during the later spread of ZIKV in the Brazil and its neighbors [61]. ZIKV has the capacity to infect the placental barrier cells and thus damage the fetus [62]. ZIKV has the sole capacity to interact with anexelekto, meaning uncontrolled (AXL) receptor, which is a tyrosine kinase receptor on the placental barrier cells while other Flaviviridae viruses cannot link with these receptors [62]. The other important arboviruses that damage the nervous system include the Japanese encephalitis virus (JEV) from the family Flaviviridae [63]. The vector for JEV is Culex mosquito species that feeds mostly on birds and livestock blood while humans are the accidental host of the virus [63, 64]. JEV causes birth defects and certain neurological complications [63]. A number of vaccines are available, but it is still endemic in South East Asian countries [63, 65]. Beside its effects on nervous system, Zika virus also damages the kidneys, eyes and genital organs [66–69].

4. Current scientific progress

For understanding the pathobiology and development of vaccines and drugs, it is important to sequence the genome and crystallize the important proteins of Zika virus. Recently, a Zika virus strain named ZikaSPH2015 genome has been fully sequenced that contained all the structural and nonstructural proteins. This strain was isolated from a patient in São Paulo, Brazil [70]. In the past, a Zika virus strain from French Polynesia was also sequenced and the results showed its Asian background [71]. It is important to compare the genome of different strains and observe the differences in them for the possible mutations. Previous phylogenetic studies showed that there are basically two strains, one Asian and the other African, with some modifications due to glycosylation of amino acid sequence [72]. Recently, Kostyuchenko et al. have resolved a 3.7 Å resolution structure of Zika virus through cryoelectron microscopy [1], while Sirohi et al. have resolved a 3.8 Å resolution structure of mature particles [73]. As compared to other flaviviruses, Zika virus is stable even at 40°C. They also showed that its envelope (E) protein is similar in structure and function to that of West Nile and Japanese encephalitis viruses and dengue virus [1].

The isolation, purification and crystallization of different Zika virus proteins are also in progress and they will be helpful in designing vaccines and chemotherapeutic agents (**Figure 1**) [74]. The nonstructural protein (NS1) crystal structure showed unique electrostatic properties at the host interaction site and it is possible in different modes of action as compared to other NS1 proteins of flaviviruses [75]. It was also observed that the NS1 codons are helping more in adaptation to humans, as in the past Zika virus was more restricted to the zoonotic cycle [76]. Tian et al. resolved the crystal structure of helicase of Zika virus and they observed that

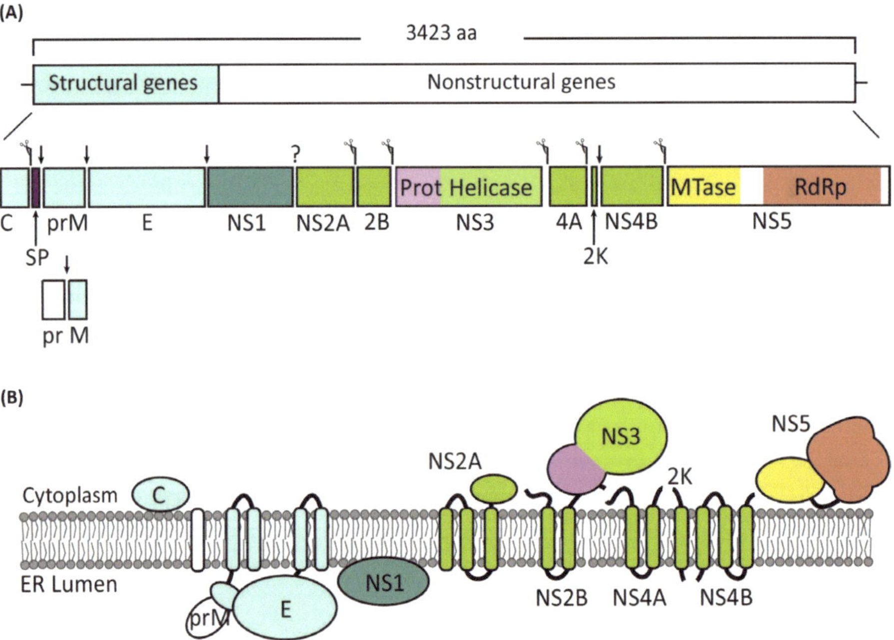

Figure 1. Sketch of the 11 kb genome of ZIKV and its important proteins. (A) The polyprotein and its cleavage products. (B) Topology of the polyprotein inside the membrane [85].

structurally it is similar to that of dengue virus but there are some differences in the motor domain [77]. That is why, Zika helicase binds to RNA differently and certain conformational changes can be seen in the motor domain [77]. Jain et al. resolved a 1.6 Å resolution structure of nonstructural (NS3) protein that acts as an RNA helicase of Zika virus, which was a French Polynesia strain [78]. This NS3 has similarity with that of dengue NS3 but there are some variations in its binding loops of ATP and RNA. This NS3 structure might be helpful in making necessary drugs to control Zika virus [78]. Rossi et al. studied Zika virus in murine models and this study could be helpful in the development of vaccines and drugs [79]. Shan et al. made an infectious cDNA clone of Zika virus using the clinical isolated strain of Zika virus of Asian origin [80]. This cDNA clone can infect the brain cells in mice and it can also infect the Aedes mosquito; thus, it is a useful tool to study the virus transmission, related diseases and research suitable therapeutics [80]. Like dengue virus, model studies suggested that there are chances that it will spread throughout the world [81, 82] and it may be possible to spread along with other flaviviruses like dengue, West Nile virus, chikungunya and Stratford virus due to the common vector Aedes [83, 84].

5. Vaccine and drugs

Some scientific groups have suggested releasing the Wolbachia-harboring mosquitoes in the environment as competitors for Aedes mosquitos because they are resistant and they do not carry Zika virus; in this way, we can control the spread of Zika virus [86]. There is no vaccine or medicine available for the treatment of Zika virus, so it is important to be safe from mosquito bite and sexual contacts in Zika virus–affected places, which is the only solution [34, 87]. The health departments of USA are giving guidelines from time to time for prevention and control of Zika virus [88, 89]. Currently, different research institutes are working in developing a vaccine against this virus, but it will certainly take time in preparing it [90–93]. It has been observed that the polyphenol epigallocatechin 3-gallate (EGCG) inhibits the entry of Zika virus into the cell [94]. This polyphenol is present in green tea and is also part of many dietary supplements. Such observations were also recorded previously for other viruses, especially hepatitis C and herpes simplex, but how efficient this compound exactly is and its mechanism of action are still not clear [95–99]. Larocca et al. made DNA vaccines that express the premembrane and envelope protein of Zika and offer protection against the Brazilian virus challenge in murine models [100]. This initial vaccine development is a positive sign of a more effective vaccine development and its marketing in the near future for the control of Zika virus. It is also reported that those parts of Brazil where yellow fever vaccination was done have lower level of microcephaly cases due to Zika virus [101]. Thus, there is a possibility that the yellow fever vaccine may be helpful for the treatment of Zika virus infection. An opposite case to this was reported where a male American traveler had yellow fever vaccination record but still he developed symptoms of Zika virus infection like rash, fever, extreme fatigue, back pain and conjunctivitis [102]. There are also suggestions to develop better mouse and other animal models that can be used to study the pathogenesis and vaccine development of Zika virus; further, the already present neuroteratogenic problem–causing virus models can also be utilized for Zika virus infection [103–106]. Computational studies predicted that the antibodies are produced in humans during Zika virus and the infection causes an autoimmune response that results in rash, microcephaly and other symptoms [107]. Therefore, remedy for such autoimmune proteins should also be considered for treatment during development of vaccines and drugs. On the drug and development side, it was noted that the 7-deaza-2′-C-methyladenosine (7DMA) has the potency to inhibit the replication of Zika virus in mice, and further trials should be performed to check its efficiency against different strains of Zika virus [108]. Eyer et al. tested a series of substituted nucleoside bases to inhibit the replication of Zika virus and they observed that the 2′-C-methylated nucleosides are promising drug candidates [109]. For research purposes, it has been suggested that the famous adenosine analog NITD008 that has been used in the past against other viruses [110–112] also has the antiviral activity both *in vitro* and *in vivo* against Zika virus [113]. Now, the sequences and in some cases the crystal structure of different important Zika virus proteins are available and they can be targeted through computational drug designing techniques [114–117]. Similarly, the already available data on the closely related viruses like dengue virus enzymes can be exploited for novel drug design [118]. In mouse model, it was noted that 25-hydroxycholesterol prevents the entry of virus particles inside the cell and is the first line of defense molecule [119]. One of the positive aspects of ZIKV is that it can be utilized for brain cancer treatment. It was observed that the ZIKV possesses oncolytic properties and it can destroy brain cancer cells called glioblastoma with high specificity [120].

6. Conclusions

The way Zika virus is spreading silently first in South America and now in Asian countries is quite alarming. If the spread continues and microcephaly cases increase, then a whole generation born might be a patient of microcephaly. Zika virus has already affected the tourism economy of different countries, and if this spread persists in Asia, then the loss will be too large. There is a need to figure out how exactly Zika virus causes microcephaly. The current research progress in terms of vaccine and drug development is at a snail pace and collaborative efforts are required to control this viral disease. Further, the spread of Aedes mosquitoes is also alarming as they are the source of other viruses too, and controlling the vector means control on ZIKV, dengue and chikungunya.

Acknowledgements

We appreciate the Deanship of Scientific Research at king Saud University for its funding to this Prolific Research group (PRG-1437-29). The authors declare no competing interests.

Author details

Syed Lal Badshah[1*†], Yahia Nasser Mabkhot[2*†], Nasir Ahmad[1], Shazia Syed[3] and Abdul Naeem[4]

*Address all correspondence to: shahbiochemist@gmail.com and yahia@ksu.edu.sa

1 Department of Chemistry, Islamia College University, Peshawar, Pakistan

2 Department of Chemistry, College of Sciences, King Saud University, Saudi Arabia

3 Department of Zoology, University of Peshawar, Peshawar, Pakistan

4 National Center of Excellence in Physical Chemistry, University of Peshawar, Peshawar, Pakistan

† Both authors contributed equally in design and discussion of this review manuscript

References

[1] Kostyuchenko VA, Lim EXY, Zhang S, Fibriansah G, Ng T-S, Ooi JSG, Shi J, Lok S-M. Structure of the thermally stable Zika virus. Nature. 2016;**533**:425-428. DOI: 10.1038/nature17994

[2] Ledermann JP, Guillaumot L, Yug L, Saweyog SC, Tided M, Machieng P, Pretrick M, Marfel M, Griggs A, Bel M, Duffy MR, Hancock WT, Ho-Chen T, Powers AM. *Aedes hensilli* as a potential vector of Chikungunya and Zika viruses. PLoS Neglected Tropical Diseases. 2014;**8**. DOI: 10.1371/journal.pntd.0003188

[3] Chang C, Ortiz K, Ansari A, Gershwin ME. The Zika outbreak of the 21st century. Journal of Autoimmunity. 2016;**68**:1-13. DOI: 10.1016/j.jaut.2016.02.006

[4] Petersen LR, Jamieson DJ, Powers AM, Honein MA. Zika Virus. The New England Journal of Medicine. 2016;**374**:1552-1563. DOI: 10.1056/NEJMra1602113

[5] Sampathkumar P, Sanchez JL. Zika virus in the Americas: A review for clinicians. Mayo Clinic Proceedings. 2016;**91**:514-521. DOI: 10.1016/j.mayocp.2016.02.017

[6] Imperato PJ. The convergence of a virus, mosquitoes, and human travel in globalizing the Zika epidemic. Journal of Community Health. 2016;**41**:674-679. DOI: 10.1007/s10900-016-0177-7

[7] Martínez de Salazar P, Suy A, Sánchez-Montalvá A, Rodó C, Salvador F, Molina I. Zika fever. Enfermedades Infecciosas y Microbiología Clínica. 2016;**34**:247-252. DOI: 10.1016/j.eimc.2016.02.016

[8] Zanluca C, Duarte dos Santos CN. Zika virus—An overview. Microbes and Infection. 2016;**18**:295-301. DOI: 10.1016/j.micinf.2016.03.003

[9] Jamil Z, Waheed Y, Durrani TZ. Zika virus, a pathway to new challenges. Asian Pacific Journal of Tropical Medicine. 2016;**9**:626-629. DOI: 10.1016/j.apjtm.2016.05.020

[10] Weinbren M, Williams M. Zika virus: Further isolations in the zika area, and some studies on the strains isolated. Transactions of the Royal Society of Tropical Medicine and Hygiene. 1958;**52**:263-268. DOI: 10.1016/0035-9203(58)90085-3

[11] Haddow AJ, Williams MC, Woodall JP, Simpson DI, Goma LK. Twelve isolations of zika virus from Aedes (Stegomyia) Africanus (Theobald) taken in and above a Uganda forest. Bulletin of the World Health Organization. 1964;**31**:57-69

[12] Kirya BG, Mukwaya LG, Sempala SDK. A yellow fever epizootic in zika forest, uganda, during 1972: Part 1: Virus isolation and sentinel monkeys. Transactions of the Royal Society of Tropical Medicine and Hygiene. 1977;**71**:254-260. DOI: 10.1016/0035-9203(77)90020-7

[13] Dick GWA. Zika virus (II). Pathogenicity and physical properties. Transactions of the Royal Society of Tropical Medicine and Hygiene. 1952;**46**:521-534. DOI: 10.1016/0035-9203(52)90043-6

[14] Wong PSJ, Li MZI, Chong CS, Ng LC, Tan CH. Aedes (Stegomyia) albopictus (Skuse): A potential vector of Zika virus in Singapore. PLoS Neglected Tropical Diseases. 2013;**7**. DOI: 10.1371/journal.pntd.0002348

[15] Simpson DIH. Zika virus infection in man. Transactions of the Royal Society of Tropical Medicine and Hygiene. 1964;**58**:339-348. DOI: 10.1016/0035-9203(64)90201-9

[16] Goorhuis A, Von Eije KJ, Douma RA, Rijnberg N, Van Vugt M, Stijnis C, Grobusch MP. Zika virus and the risk of imported infection in returned travelers: Implications for clinical care. Travel Medicine and Infectious Disease. 2016;**14**:13-15. DOI: 10.1016/j.tmaid.2016.01.008

[17] Ioos S, Mallet HP, Leparc Goffart I, Gauthier V, Cardoso T, Herida M. Current Zika virus epidemiology and recent epidemics. Médecine et Maladies Infectieuses. 2014;**44**:302-307. DOI: 10.1016/j.medmal.2014.04.008

[18] Gyurech D, Schilling J, Schmidt-Chanasit J, Cassinotti P, Kaeppeli F, Dobec M. False positive dengue NS1 antigen test in a traveller with an acute Zika virus infection imported into Switzerland. Swiss Medical Weekly. 2016;**146**:w14296. DOI: 10.4414/smw.2016.14296

[19] Basarab M, Bowman C, Aarons EJ, Cropley I. Zika virus. BMJ. 2016;**1049**:i1049. DOI: 10.1136/bmj.i1049

[20] Gatherer D, Kohl A. Zika virus: A previously slow pandemic spreads rapidly through the Americas. The Journal of General Virology. 2016;**97**:269-273. DOI: 10.1099/jgv.0.000381

[21] Lucey DR, Gostin LO. The emerging Zika pandemic: Enhancing preparedness. Journal of the American Medical Association. 2016;**315**:865-866. DOI: 10.1001/jama.2016.0904

[22] Darwish MA, Hoogstraal H, Roberts TJ, Ahmed IP, Omar F. A sero-epidemiological survey for certain arboviruses (Togaviridae) in Pakistan. Transactions of the Royal Society of Tropical Medicine and Hygiene. 1983;**77**:442-445. DOI: 10.1016/0035-9203(83)90106-2

[23] Bhide P, Kar A. Birth prevalence of microcephaly in India. Bulletin of the World Health Organization. 2016:1-11. DOI: 10.2471/BLT.16.172080

[24] Shinohara K, Kutsuna S, Takasaki T, Moi ML, Ikeda M, Kotaki A, Yamamoto K, Fujiya Y, Mawatari M, Takeshita N, Hayakawa K, Kanagawa S, Kato Y, Ohmagari N. Zika fever imported from Thailand to Japan, and diagnosed by PCR in the urines. Journal of Travel Medicine. 2016;**23**:tav011. DOI: 10.1093/jtm/tav011

[25] Heang V, Yasuda CY, Sovann L, Haddow AD, da Rosa APT, Tesh RB, Kasper MR. Zika virus infection, Cambodia, 2010. Emerging Infectious Diseases. 2012;**18**:349-351. DOI: 10.3201/eid1802.111224

[26] Diallo D, Sall AA, Diagne CT, Faye O, Faye O, Ba Y, Hanley KA, Buenemann M, Weaver SC, Diallo M. Zika virus emergence in mosquitoes in Southeastern Senegal, 2011. PLoS One. 2014;**9**. DOI: 10.1371/journal.pone.0109442

[27] Diagne CT, Diallo D, Faye O, Ba Y, Faye O, Gaye A, Dia I, Weaver SC, Sall AA, Diallo M. Potential of selected Senegalese Aedes spp. mosquitoes (Diptera: Culicidae) to transmit Zika virus. BMC Infectious Diseases. 2015;**15**:492. DOI: 10.1186/s12879-015-1231-2

[28] Chan JFW, Choi GKY, Yip CCY, Cheng VCC, Yuen KY. Zika fever and congenital Zika syndrome: An unexpected emerging arboviral disease. The Journal of Infection. 2016;**72**:507-524. DOI: 10.1016/j.jinf.2016.02.011

[29] Marcondes CB, Ximenes, De FF, De M. Zika virus in Brazil and the danger of infestation by Aedes (Stegomyia) mosquitoes. Revista da Sociedade Brasileira de Medicina Tropical. 2015;**49**:4-10. DOI: 10.1590/0037-8682-0220-2015

[30] Lancet T. Zika virus: A new global threat for 2016. Lancet. 2016;**387**:96. DOI: 10.1016/S0140-6736(16)00014-3

[31] Zanluca C, de Melo VCA, Mosimann ALP, dos Santos GIV, dos Santos CND, Luz K. First report of autochthonous transmission of Zika virus in Brazil. Memórias do Instituto Oswaldo Cruz. 2015;**110**:569-572. DOI: 10.1590/0074-02760150192

[32] Hamel R, Dejarnac O, Wichit S, Ekchariyawat P, Neyret A, Luplertlop N, Perera-Lecoin M, Surasombatpattana P, Talignani L, Thomas F, Cao-Lormeau V-M, Choumet V, Briant L, Desprès P, Amara A, Yssel H, Missé D. Biology of Zika virus infection in human skin cells. Journal of Virology. 2015;**89**:8880-8896. DOI: 10.1128/JVI.00354-15

[33] Colt S, Garcia-Casal MN, Peña-Rosas JP, Finkelstein JL, Rayco-Solon P, Weise Prinzo ZC, Mehta S. Transmission of Zika virus through breast milk and other breastfeeding-related bodily-fluids: A systematic review. PLoS Neglected Tropical Diseases. 2017;**11**. DOI: 10.1371/journal.pntd.0005528

[34] Oster AM, Brooks JT, Stryker JE, Kachur RE, Mead P, Pesik NT, Petersen LR. Interim guidelines for prevention of sexual transmission of Zika virus—United States, 2016. MMWR. Morbidity and Mortality Weekly Report. 2016;**65**:120-121. DOI: 10.15585/mmwr.mm6505e1

[35] Musso D, Roche C, Robin E, Nhan T, Teissier A, Cao-Lormeau VM. Potential sexual transmission of zika virus. Emerging Infectious Diseases. 2015;**21**:359-361. DOI: 10.3201/eid2102.141363

[36] European Centre for Disease Prevention and Control. Zika virus disease epidemic: Potential association with microcephaly and Guillain–Barré syndrome. Rapid Risk Assessment. 2016. http://ecdc.europa.eu/en/publications/Publications/zika-virus-rapid-risk-assessment-8-february-2016.pdf

[37] Barton MA, Salvadori MI. Zika virus and microcephaly. CMAJ. 2016;**188**:e118-e119. DOI: 10.1503/cmaj.160179

[38] Rasmussen SA, Jamieson DJ, Honein MA, Petersen LR. Zika virus and birth defects—Reviewing the evidence for causality. The New England Journal of Medicine. 2016;**374**:1981-1987. DOI: 10.1056/NEJMsr1604338

[39] Mlakar J, Korva M, Tul N, Popović M, Poljšak-Prijatelj M, Mraz J, Kolenc M, Rus K, Vipotnik T, Vodušek V, Vizjak A, Pižem J, Petrovec M, Županc T. Zika virus associated with microcephaly. The New England Journal of Medicine. 2016;**374**:951-958. DOI: 10.1056/NEJMoa1600651

[40] Michael PD, Johansson A, Mier-y-Teran-Romero L, Reefhuis J, Gilboa SM, Hills SL, Mlakar J, Korva M, Tul N, Popović M, Poljšak-Prijatelj M, Mraz J, Kolenc M, Resman Rus K, Vesnaver Vipotnik T, Fabjan Vodušek V, Vizjak A, Pižem J, Petrovec M, Avšič Županc T, Rasmussen SA, Jamieson DJ, Honein MA, Petersen LR. Zika virus associated with microcephaly. New England Journal of Medicine. 2016;**374**:1-3. DOI: 10.1056/NEJMsr1604338

[41] Cunha AJ, Magalhães-Barbosa MC, Lima-Setta F, Prata-Barbosa A. Evolution of cases of microcephaly and neurological abnormalities suggestive of congenital infection in Brazil: 2015-2016. Bulletin of the World Health Organization. 2016:1-15. DOI: 10.2471/BLT.16.173583

[42] Tang BL. Zika virus as a causative agent for primary microencephaly: The evidence so far. Archives of Microbiology. 2016;**198**:595-601. DOI: 10.1007/s00203-016-1268-7

[43] Garcez PP, Nascimento JM, de Vasconcelos JM, Madeiro da Costa R, Delvecchio R, Trindade P, Loiola EC, Higa LM, Cassoli JS, Vitória G, Sequeira PC, Sochacki J, Aguiar RS, Fuzii HT, de Filippis AMB, da Silva Gonçalves Vianez Júnior JL, Tanuri A, Martins-de-Souza D, Rehen SK. Zika virus disrupts molecular fingerprinting of human neurospheres. Scientific Reports. 2017;**7**:40780. DOI: 10.1038/srep40780

[44] Paixão ES, Barreto F, Da Glória Teixeira M, Da Conceição M, Costa N, Rodrigues LC. History, epidemiology, and clinical manifestations of Zika: A systematic review. American Journal of Public Health. 2016;**106**:606-612. DOI: 10.2105/AJPH.2016.303112

[45] Oliveira Melo AS, Malinger G, Ximenes R, Szejnfeld PO, Alves Sampaio S, Bispo De Filippis AM. Zika virus intrauterine infection causes fetal brain abnormality and microcephaly: Tip of the iceberg? Ultrasound in Obstetrics & Gynecology. 2016;**47**:6-7. DOI: 10.1002/uog.15831

[46] Mlakar J, Korva M, Tul N, Popović M, Poljšak-Prijatelj M, Mraz J, Kolenc M, Resman Rus K, Vesnaver Vipotnik T, Fabjan Vodušek V, Vizjak A, Pižem J, Petrovec M, Avšič Županc T. Zika virus associated with microcephaly. The New England Journal of Medicine. 2016;**10**:951-958. DOI: 10.1056/NEJMoa1600651

[47] Duffy MR, Chen T-H, Hancock WT, Powers AM, Kool JL, Lanciotti RS, Pretrick M, Marfel M, Holzbauer S, Dubray C, Guillaumot L, Griggs A, Bel M, Lambert AJ, Laven J, Kosoy O, Panella A, Biggerstaff BJ, Fischer M, Hayes EB. Zika virus outbreak on Yap Island, Federated States of Micronesia. The New England Journal of Medicine. 2009;**360**:2536-2543. DOI: 10.1056/NEJMoa0805715

[48] Schuler-Faccini L, Sanseverino MTV, Vianna FSL, da Silva AA, Larrandaburu M, Pereira CM, Abeche AM. Zika virus: A new human teratogen? Implications for women reproductive age. Clinical Pharmacology and Therapeutics. 2016;**100**:28-30. DOI: 10.1002/cpt.386

[49] Sarno M, Sacramento GA, Khouri R, do Rosário MS, Costa F, Archanjo G, Santos LA, Nery N, Vasilakis N, Ko AI, de Almeida ARP. Zika virus infection and stillbirths: A case of hydrops fetalis, hydranencephaly and fetal demise. PLoS Neglected Tropical Diseases. 2016;**10**. DOI: 10.1371/journal.pntd.0004517

[50] Calvet G, Aguiar RS, Melo ASO, Sampaio SA, de Filippis I, Fabri A, Araujo ESM, de Sequeira PC, de Mendonça MCL, de Oliveira L, Tschoeke DA, Schrago CG, Thompson FL, Brasil P, dos Santos FB, Nogueira RMR, Tanuri A, de Filippis AMB. Detection and sequencing of Zika virus from amniotic fluid of fetuses with microcephaly in Brazil: A case study. The Lancet Infectious Diseases. 2016;**16**:653-660. DOI: 10.1016/S1473-3099(16)00095-5

[51] Krauer F, Riesen M, Reveiz L, Oladapo OT, Martínez-Vega R, Porgo TV, Haefliger A, Broutet NJ, Low N. Zika virus infection as a cause of congenital brain abnormalities and Guillain–Barré syndrome: Systematic review. PLoS Medicine. 2017;**14**. DOI: 10.1371/journal.pmed.1002203

[52] Garcez PP, Loiola EC, Madeiro da Costa R, Higa LM, Trindade P, Delvecchio R, Nascimento JM, Brindeiro R, Tanuri A, Rehen SK. Zika virus impairs growth in human neurospheres and brain organoids. Science. 2016:aaf6116. DOI: 10.1126/science.aaf6116

[53] Vermillion MS, Lei J, Shabi Y, Baxter VK, Crilly NP, McLane M, Griffin DE, Pekosz A, Klein SL, Burd I. Intrauterine Zika virus infection of pregnant immunocompetent mice models transplacental transmission and adverse perinatal outcomes. Nature Communications. 2017;**8**:14575. DOI: 10.1038/ncomms14575

[54] Bullerdiek J, Dotzauer A, Bauer I. The mitotic spindle: Linking teratogenic effects of Zika virus with human genetics? Molecular Cytogenetics. 2016;**9**:32. DOI: 10.1186/s13039-016-0240-1

[55] Cauchemez S, Besnard M, Bompard P, Dub T, Guillemette-Artur P, Eyrolle-Guignot D, Salje H, Van Kerkhove MD, Abadie V, Garel C, Fontanet A, Mallet HP. Association between Zika virus and microcephaly in French Polynesia, 2013-15: A retrospective study. Lancet. 2016;**387**:2125-2132. DOI: 10.1016/S0140-6736(16)00651-6

[56] Panchaud A, Stojanov M, Ammerdorffer A, Vouga M, Baud D. Emerging role of Zika virus in adverse fetal and neonatal outcomes. Clinical Microbiology Reviews. 2016;**29**:659-694. DOI: 10.1128/CMR.00014-16

[57] Evans D, Nijhout F, Parens R, Morales AJ, Bar-Yam Y. A possible link between pyriproxyfen and microcephaly. bioRxiv. 2016:48538. DOI: 10.1101/048538

[58] Dzieciolowska S, Larroque A-L, Kranjec E-A, Drapeau P, Samarut E. The larvicide pyriproxyfen blamed during the Zika virus outbreak does not cause microcephaly in zebrafish embryos. Scientific Reports. 2017;**7**:40067. DOI: 10.1038/srep40067

[59] Cugola FR, Fernandes IR, Russo FB, Freitas BC, Dias JLM, Guimarães KP, Benazzato C, Almeida N, Pignatari GC, Romero S, Polonio CM, Cunha I, Freitas CL, Brandão WN, Rossato C, Andrade DG, Faria DDP, Garcez AT, Buchpigel CA, Braconi CT, Mendes E, Sall AA, Zanotto PM d A, Peron JPS, Muotri AR, Beltrão-Braga PCB. The Brazilian Zika virus strain causes birth defects in experimental models. Nature. 2016;**534**:267-271. DOI: 10.1038/nature18296

[60] Onorati M, Li Z, Liu F, Horvath TL, Lindenbach BD, Sousa AMM, Nakagawa N, Li M, Dell 'anno MT, Gulden FO, Pochareddy S, Tebbenkamp ATN, Han W, Pletikos M, Gao T, Zhu Y, Bichsel C, Varela L, Szigeti-Buck K, Lisgo S, Zhang Y, Testen A, Gao X-B, Mlakar J, Popovic M, Flamand M, Strittmatter SM, Kaczmarek LK, Anton ES, Sestan N. Zika virus disrupts Phospho-TBK1 localization and mitosis in human neuroepithelial stem cells and radial glia. Cell Reports. 2016;**16**:2576-2592. DOI: 10.1016/j.celrep.2016.08.038

[61] Yuan L. A single mutation in the prM protein of Zika virus contributes to fetal microcephaly. Science (80-.). 2017;**7120**:eaam7120. DOI: 10.1126/7120

[62] Richard AS, Shim B-S, Kwon Y-C, Zhang R, Otsuka Y, Schmitt K, Berri F, Diamond MS, Choe H. AXL-dependent infection of human fetal endothelial cells distinguishes Zika virus from other pathogenic flaviviruses. Proceedings of the National Academy of Sciences. 2017;**114**:2024-2029. DOI: 10.1073/pnas.1620558114

[63] Mansfield KL, Hernández-Triana LM, Banyard AC, Fooks AR, Johnson N. Japanese encephalitis virus infection, diagnosis and control in domestic animals. Veterinary Microbiology. 2017;**201**:85-92. DOI: 10.1016/j.vetmic.2017.01.014

[64] Duong V, Choeung R, Gorman C, Laurent D, Crabol Y, Mey C, Peng B, Di Francesco J, Hul V, Sothy H, Santy K, Richner B, Pommier J-D, Sorn S, Chevalier V, Buchy P, de Lamballerie X, Cappelle J, Horwood PF, Dussart P. Isolation and full-genome sequences of Japanese encephalitis virus genotype I strains from Cambodian human patients, mosquitoes and pigs. The Journal of General Virology. 2017;**98**:2287-2296. DOI: 10.1099/jgv.0.000892

[65] García-Nicolás O, Ricklin ME, Liniger M, Vielle NJ, Python S, Souque P, Charneau P, Summerfield A. A Japanese encephalitis virus vaccine inducing antibodies strongly enhancing in vitro infection is protective in pigs. Virus. 2017;**9**:124. DOI: 10.3390/v9050124

[66] Roach T, Alcendor DJ. Zika virus infection of cellular components of the blood-retinal barriers: Implications for viral associated congenital ocular disease. Journal of Neuroinflammation. 2017;**14**:43. DOI: 10.1186/s12974-017-0824-7

[67] Singh PK, Guest J-M, Kanwar M, Boss J, Gao N, Juzych MS, Abrams GW, Yu F-S, Kumar A. Zika virus infects cells lining the blood-retinal barrier and causes chorioretinal atrophy in mouse eyes. JCI Insight. 2017;**2**:e92340. DOI: 10.1172/jci.insight.92340

[68] de Paula Freitas B, de Oliveira Dias JRJ, Prazeres J, Sacramento GAG, Ko AI, Maia MM, Belfort RJ. Ocular findings in infants with microcephaly associated with presumed Zika virus congenital. JAMA Ophthalmology. 2016;**134**:529-535. DOI: 10.1001/jamaophthalmol.2016.0267

[69] Uraki R, Hwang J, Jurado KA, Householder S, Yockey LJ, Hastings AK, Homer RJ, Iwasaki A, Fikrig E. Zika virus causes testicular atrophy. Science Advances. 2017;**3**: e1602899. DOI: 10.1126/sciadv.1602899

[70] Cunha MS, Esposito DLA, Rocco IM, Maeda AY, Vasami FGS, Nogueira JS, de Souza RP, Suzuki A, Addas-Carvalho M, Barjas-Castro M d L, Resende MR, Stucchi RSB, Boin I d FSF, Katz G, Angerami RN, da Fonseca BAL. First complete genome sequence of Zika virus (Flaviviridae, Flavivirus) from an autochthonous transmission in Brazil. Genome Announcements. 2016;**4**:2015-2016. DOI: 10.1128/genomeA.00032-16

[71] Baronti C, Piorkowski G, Charrel RN, Boubis L, Leparc-Goffart I, de Lamballerie X. Complete coding sequence of zika virus from a French polynesia outbreak in 2013. Genome Announcements. 2014 pii: e00500-14;**2**. DOI: 10.1128/genomeA.00500-14

[72] Haddow AD, Schuh AJ, Yasuda CY, Kasper MR, Heang V, Huy R, Guzman H, Tesh RB, Weaver SC. Genetic characterization of zika virus strains: Geographic expansion of the asian lineage. PLoS Neglected Tropical Diseases. 2012;**6**. DOI: 10.1371/journal.pntd.0001477

[73] Sirohi D, Chen Z, Sun L, Klose T, Pierson TC, Rossmann MG, Kuhn RJ. The 3.8 Å resolution cryo-EM structure of Zika virus. Science (80-). 2016;**5316**:1-7. DOI: 10.1126/science.aaf5316

[74] Badshah S, Naeem A, Mabkhot Y. The new high resolution crystal structure of NS2B-NS3 protease of Zika virus. Virus. 2017;**9**:7. DOI: 10.3390/v9010007

[75] Song H, Qi J, Haywood J, Shi Y, Gao GF. Zika virus NS1 structure reveals diversity of electrostatic surfaces among flaviviruses. Nature Structural & Molecular Biology. 2016;**23**(5):456-458. DOI: 10.1038/nsmb.3213

[76] Freire CCM, Iamarino A, Neto DFL, Sall AA, Marinho P, Zanotto A. Spread of the pandemic Zika virus lineage is associated with NS1 codon usage adaptation in humans. bioRxiv. 2015:1-8. DOI: 10.1101/032839

[77] Tian H, Ji X, Yang X, Zhang Z, Lu Z, Yang K, Chen C, Zhao Q, Chi H, Mu Z, Xie W, Wang Z, Lou H, Yang H, Rao Z. Structural basis of Zika virus helicase in recognizing its substrates. Protein & Cell. 2016;**7**:562-570. DOI: 10.1007/s13238-016-0293-2

[78] Jain R, Coloma J, Garcia-Sastre A, Aggarwal AK. Structure of the NS3 helicase from Zika virus. Nature Structural & Molecular Biology. 2016;**23**:752-754. DOI: 10.1038/nsmb.3258

[79] Rossi SL, Tesh RB, Azar SR, Muruato AE, Hanley KA, Auguste AJ, Langsjoen RM, Paessler S, Vasilakis N, Weaver SC. Characterization of a novel murine model to study Zika virus. American Journal of Tropical Medicine and Hygiene. 2016;**94**:1362-1369. DOI: 10.4269/ajtmh.16-0111

[80] Shan C, Xie X, Muruato AE, Rossi SL, Roundy CM, Azar SR, Yang Y, Tesh RB, Bourne N, Barrett AD, Vasilakis N, Weaver SC, Shi PY. An infectious cDNA clone of Zika virus to study viral virulence, mosquito transmission, and antiviral inhibitors. Cell Host & Microbe. 2016;**19**:891-900. DOI: 10.1016/j.chom.2016.05.004

[81] Bogoch II, Brady OJ, Kraemer MUG, German M, Creatore MI, Kulkarni MA, Brownstein JS, Mekaru SR, Hay SI, Groot E, Watts A, Khan K. Anticipating the international spread of Zika virus from Brazil. Lancet. 2016;**387**:335-336. DOI: 10.1016/S0140-6736(16)00080-5

[82] Hayes EB. Zika virus outside Africa. Emerging Infectious Diseases. 2009;**15**:1347-1350. DOI: 10.3201/eid1509.090442

[83] Attar N. ZIKA virus circulates in new regions. Nature Reviews. Microbiology. 2016;**14**:62. DOI: 10.1038/nrmicro.2015.28

[84] Toi CS, Webb CE, Haniotis J, Clancy J, Doggett SL. Seasonal activity, vector relationships and genetic analysis of mosquito-borne Stratford virus. PLoS One. 2017;**12**. DOI: 10.1371/journal.pone.0173105

[85] Shi Y, Gao GF. Structural biology of the Zika virus. Trends in Biochemical Sciences. 2017;**42**:443-456. DOI: 10.1016/j.tibs.2017.02.009

[86] Leandro H, Dutra C, Rocha MN, Braga F, Dias S, Mansur SB, Caragata EP, Moreira LA. Wolbachia blocks currently circulating Zika virus isolates in Brazilian *Aedes aegypti* mosquitoes. Cell Host & Microbe. 2016;**19**:1-4. DOI: 10.1016/j.chom.2016.04.021

[87] Ther JCB, Zhao Y. Challenges in searching for Zika therapeutics. Journal of Chemical Biology & Therapeutics. 2016;**2**:1-2. DOI: 10.4172/jcbt.1000e102

[88] Staples JE, Dziuban EJ, Fischer M, Cragan JD, Rasmussen SA, Cannon MJ, Frey MT, Renquist CM, Lanciotti RS, Muñoz JL, Powers AM, Honein MA, Moore CA, Munoz JL, Powers AM, Honein MA, Moore CA. Interim guidelines for the evaluation and testing of infants with possible congenital Zika virus infection—United States, 2016. MMWR. Morbidity and Mortality Weekly Report. 2016;**65**:63-67. DOI: 10.15585/mmwr.mm6503e3

[89] Petersen EE, Staples JE, Meaney-delman D, Fischer M, Ellington SR, Callaghan WM, Jamieson DJ. Interim guidelines for pregnant women during a Zika virus outbreak—United States, 2016. MMWR. Morbidity and Mortality Weekly Report. 2016;**65**:30-33. DOI: 10.15585/mmwr.mm6502e1

[90] Cohen J. Infectious disease. The race for a Zika vaccine is on. Science (80-). 2016;**351**:543-544. DOI: 10.1126/science.351.6273.543

[91] Check Hayden E. The race is on to develop Zika vaccine. Nature. 2016. DOI: 10.1038/nature.2016.19634

[92] Shan C, Xie X, Barrett ADT, Garcia-Blanco MA, Tesh RB, Vasconcelos PFDC, Vasilakis N, Weaver SC, Shi PY. Zika virus: Diagnosis, therapeutics, and vaccine. ACS Infectious Diseases. 2016;**2**:170-172. DOI: 10.1021/acsinfecdis.6b00030

[93] Dawes BE, Smalley CA, Tiner BL, Beasley DW, Milligan GN, Reece LM, Hombach J, Barrett AD. Research and development of Zika virus vaccines. Npj Vaccines. 2016;**1**:16007. DOI: 10.1038/npjvaccines.2016.7

[94] Carneiro BM, Batista MN, Braga ACS, Nogueira ML, Rahal P. The green tea molecule EGCG inhibits Zika virus entry. Virology. 2016;**496**:215-218. DOI: 10.1016/j.virol.2016.06.012

[95] Isaacs CE, Wen GY, Xu W, Jia JH, Rohan L, Corbo C, Di Maggio V, Jenkins EC, Hillier S. Epigallocatechin gallate inactivates clinical isolates of herpes simplex virus. Antimicrobial Agents and Chemotherapy. 2008;**52**:962-970. DOI: 10.1128/AAC.00825-07

[96] Ciesek S, von Hahn T, Colpitts CC, Schang LM, Friesland M, Steinmann J, Manns MP, Ott M, Wedemeyer H, Meuleman P, Pietschmann T, Steinmann E. The green tea polyphenol, epigallocatechin-3-gallate, inhibits hepatitis C virus entry. Hepatology. 2011;**54**:1947-1955. DOI: 10.1002/hep.24610

[97] De Oliveira A, Adams SD, Lee LH, Murray SR, Hsu SD, Hammond JR, Dickinson D, Chen P, Chu TC. Inhibition of herpes simplex virus type 1 with the modified green tea polyphenol palmitoyl-epigallocatechin gallate. Food and Chemical Toxicology. 2013;**52**:207-215. DOI: 10.1016/j.fct.2012.11.006

[98] Dhiman RK. The green tea polyphenol, epigallocatechin-3-gallate (EGCG)-one step forward in antiviral therapy against hepatitis C virus. Journal of Clinical and Experimental Hepatology. 2011;**1**:159-160. DOI: 10.1016/S0973-6883(11)60232-6

[99] Chen C, Qiu H, Gong J, Liu Q, Xiao H, Chen XW, Sun BL, Yang RG. (−)-Epigallocatechin-3-gallate inhibits the replication cycle of hepatitis C virus. Archives of Virology. 2012;**157**:1301-1312. DOI: 10.1007/s00705-012-1304-0

[100] Larocca RA, Abbink P, Peron JPS, Zanotto PM, de A, Iampietro MJ, Badamchi-Zadeh A, Boyd M, Ng'ang'a D, Kirilova M, Nityanandam R, Mercado NB, Li Z, Moseley ET, Bricault CA, Borducchi EN, Giglio PB, Jetton D, Neubauer G, Nkolola JP, Maxfield LF, La Barrera RAD, Jarman RG, Eckels KH, Michael NL, Thomas SJ, Barouch DH. Vaccine protection against Zika virus from Brazil. Nature. 2016;**536**:474-478. DOI: 10.1038/nature18952

[101] De Góes Cavalcanti LP, Tauil PL, Alencar CH, Oliveira W, Teixeira MM, Heukelbach J. Zika virus infection, associated microcephaly, and low yellow fever vaccination coverage in Brazil: Is there any causal link. Journal of Infection in Developing Countries. 2016;**10**:563-566. DOI: 10.3855/jidc.8575

[102] Summers DJ, Acosta RW, Acosta AM. Zika virus in an American recreational Traveler. Journal of Travel Medicine. 2015;**22**:338-340. DOI: 10.1111/jtm.12208

[103] Lazear HM, Govero J, Smith AM, Platt DJ, Fernandez E, Miner JJ, Correspondence MSD, Diamond MS. A mouse model of Zika virus pathogenesis. Cell Host & Microbe. 2016;**19**:1-11. DOI: 10.1016/j.chom.2016.03.010

[104] Shah A, Kumar A. Zika virus infection and development of a murine model. Neurotoxicity Research. 2016;**30**:131-134. DOI: 10.1007/s12640-016-9635-3

[105] Kim K, Shresta S. Neuroteratogenic viruses and lessons for Zika virus models. Trends in Microbiology. 2016;**24**(8):622-636. DOI: 10.1016/j.tim.2016.06.002

[106] Bardina SV, Bunduc P, Tripathi S, Duehr J, Frere JJ, Brown JA, Nachbagauer R, Foster GA, Krysztof D, Tortorella D, Stramer SL, García-Sastre A, Krammer F, Lim JK. Enhancement of Zika virus pathogenesis by preexisting antiflavivirus immunity. Science (80-). 2017;**356**: 175-180. DOI: 10.1126/science.aal4365

[107] Homan J, Malone RW, Darnell SJ, Bremel RD. Antibody mediated epitope mimicry in the pathogenesis of Zika virus related disease. bioRxiv. 2016:44834. DOI: 10.1101/044834

[108] Zmurko J, Marques RE, Schols D, Verbeken E, Kaptein SJF, Neyts J. The viral polymerase inhibitor 7-Deaza-2′-C-methyladenosine is a potent inhibitor of in vitro Zika virus replication and delays disease progression in a robust mouse infection model. PLoS Neglected Tropical Diseases. 2016;**10**. DOI: 10.1371/journal.pntd.0004695

[109] Eyer L, Nencka R, Huvarová I, Palus M, Joao Alves M, Gould EA, De Clercq E, Růžek D. Nucleoside inhibitors of Zika virus. The Journal of Infectious Diseases. 2016;**214**(5):707-711. DOI: 10.1093/infdis/jiw226

[110] Qing J, Luo R, Wang Y, Nong J, Wu M, Shao Y, Tang R, Yu X, Yin Z, Sun Y. Resistance analysis and characterization of NITD008 as an adenosine analog inhibitor against hepatitis C virus. Antiviral Research. 2016;**126**:43-54. DOI: 10.1016/j.antiviral.2015.12.010

[111] Yin Z, Chen Y-L, Schul W, Wang Q-Y, Gu F, Duraiswamy J, Kondreddi RR, Niyomrattanakit P, Lakshminarayana SB, Goh A, Xu HY, Liu W, Liu B, Lim JYH, Ng CY, Qing M, Lim CC, Yip A, Wang G, Chan WL, Tan HP, Lin K, Zhang B, Zou G, Bernard KA, Garrett C, Beltz K, Dong M, Weaver M, He H, Pichota A, Dartois V, Keller TH, Shi P-Y. An adenosine nucleoside inhibitor of dengue virus. Proceedings of the National Academy of Sciences of the United States of America. 2009;**106**:20435-20439. DOI: 10.1073/pnas.0907010106

[112] Deng C-L, Yeo H, Ye H-Q, Liu S-Q, Shang B-D, Gong P, Alonso S, Shi P-Y, Zhang B. Inhibition of enterovirus 71 by adenosine analog NITD008. Journal of Virology. 2014;**88**:11915-11923. DOI: 10.1128/JVI.01207-14

[113] Deng Y-Q, Zhang N-N, Li C-F, Tian M, Hao J-N, Xie X-P, Shi P-Y, Qin C-F. Adenosine analog NITD008 is a potent inhibitor of Zika virus. Open Forum Infectious Diseases. 2016;**3**. DOI: 10.1093/ofid/ofw175

[114] Carrasco JPC, Coronado Parra T. Application of computational drug discovery techniques for designing new drugs against Zika virus, drug des. Open Access. 2016;**5**. DOI: 10.4172/2169-0138.1000e131

[115] Ekins S, Mietchen D, Coffee M, Stratton TP, Freundlich JS, Freitas-Junior L, Muratov E, Siqueira-Neto J, Williams AJ, Andrade C. Open drug discovery for the Zika virus. F1000Research. 2016;**5**:150. DOI: 10.12688/f1000research.8013.1

[116] Ekins S, Liebler J, Neves BJ, Lewis WG, Coffee M, Bienstock R, Southan C, Andrade CH. Illustrating and homology modeling the proteins of the Zika virus. F1000Research. 2016;**5**:275. DOI: 10.12688/f1000research.8213.1

[117] Ramharack P, Soliman MES, Faye O, Freire CCM, Iamarino A, Faye O, de Oliveira JVC, Diallo M, Bhakat S, Karubiu W, Jayaprakash V, Soliman MES, Chen LH, Hamer DH, Campos GS, Bandeira AC, Sardi SI, Mahfuz M, Khan A, Al MH, Hasan M, Parvin A, Rahman N, Roa M, WHO, Musso D, Roche C, Robin E, Nhan T, Teissier A, Cao-Lormeau VM, Ekins S, Mietchen D, Coffee M, Stratton T, Freundlich J, Freitas-Junior L, Singh RK, Dhama K, Malik YS, Ramakrishnan MA, Karthik K, Tiwari R, Lissauer D, Smit E, Kilby MD, Pettersen EF, Goddard TD, Huang CC, Couch GS, Greenblatt DM, Meng EC, Panchaud A, Stojanov M, Ammerdorffer A, Vouga M, Chibueze EC, Tirado V, Olukunmi O, Mlakar J, Korva M, Tul N, Popović M, Poljšak-Prijatelj M, Mraz J, Turmel JM, Amgueguen P, Hubert B, Vandamme YM, Maquart M, Guillou-Guillemette HL, Art IL-G, Melo ASO, Malinger G, Ximenes R, Szejnfeld PO, Sampaio SA, De Filippis AMB, Plourde AR, Bloch EM, Calvet G, Aguiarv RS, Melo AS, Sampaio SA, de Filippis I, Fabri A, Petersen LR, Jamieson DJ, Powers AM, Honein MA, Waggoner JJ, Pinsky BA, Gourinat AC, Connor O, Calvez E, Goarant C, Dupont-Rouzeyrol M, Charrel RN, Leparc-Goffart I, Pas S, de Lamballerie X, Koopmans M, Reuskens C, Haddow AD, Schuh AJ, Yasuda CY, Kasper MR, Heang V, Huy R, Hayes EB, Dick GWA, Kitchen SF, Haddow AJ, Duffy MR, Chen T-H, Hancock WT, Powers AM, Kool JL, Lanciotti RS, Malone RW, Homan J, Callahan MV, Glasspool-Malone J, Damodaran L, Schneider ADB, Basarab M, Bowman C, Aarons EJ, Cropley I, Saiz J-C, Vázquez-Calvo Á, Blázquez AB, Merino-Ramos T, Escribano-Romero E, Martín-Acebes MA, Ekins S, Liebler J, Neves BJ, Lewis WG, Coffee M, Bienstock R, Hughes JP, Rees SS, Kalindjian SB, Philpott KL, Noble CG, Chen YL, Dong H, Gu F, Lim SP, Schul W, Perera R, Khaliq M, Kuhn RJ, Chappell KJ, Stoermer MJ, Fairlie DP, Young PR, Agnihotri S, Narula R, Joshi K, Rana S, Singh M, Tian H, Ji X, Yang X, Xie W, Yang K, Chen C, Song H, Qi J, Haywood J, Shi Y, Gao GF, Youn S, Ambrose RL, Mackenzie JM, Diamond MS, Beatty PR, Puerta-Guardo H, Killingbeck SS, Glasner DR, Hopkins K, Harris E, Stahla-Beek HJ, April DG, Saeedi BJ, Hannah AM, Keenan SM, Geiss BJ, Sirohi D, Chen Z, Sun L, Klose T, Pierson TC, Rossmann MG, Courageot M, Frenkiel M, Dos SD, Deubel V, Desprès P, Duarte C, Hamel R, Dejarnac O, Wichit S, Ekchariyawat P, Neyret A, Luplertlop N, Noppakunmongkolchai W, Poyomtip T, Jittawuttipoka T, Luplertlop N, Sakuntabhai A, Chimnaronk S, Hirsch AJ, Zhang R, Miner JJ, Gorman MJ, Rausch K, Ramage H, White JP, Mahfuz M, Ali K, Mahmud HA, Mahmudul H, Parvin A, Nazibur R, Badier SM, Kostyuchenko VA, Lim EXY, Zhang S, Fibriansah G, Ng T-S, Ooi JSG, Dai L, Song J, Lu X, Deng Y-Q, Musyoki AM, Cheng H, Barba-Spaeth G, Dejnirattisai W, Rouvinski A, Vaney M-C, Medits I, Sharma A, Weltman JK, Mandal S, Moudgil M, Mandal SK, Song CM, Lim SJ, Tong JC, Bamborough P, Cohen FE, Anderson AC, Huang HJ, Yu HW, Chen CY, Hsu CH, Chen HY, Lee KJ, Honarparvar B, Govender T, Maguire GEM, Soliman MES, Kruger HG, Hooft RW, Sander C, Vriend G, Chetty S, Soliman MES, Fathi SMS, Tuszynski JA, Lionta E, Spyrou G, Vassilatis DK, Cournia Z, Ramesh M, Vepuri SB, Oosthuizen F, Soliman ME, Reddy SS, Pati PP, Kumar PP,

Pradeep H, Sastry NN, Sliwoski G, Kothiwale S, Meiler J, Lowe EW, Kumalo HM, Soliman ME, Delvecchio R, Higa LM, Pezzuto P, Valadao AL, Garcez PP, Monteiro FL, Dowall SD, Graham VA, Rayner E, Atkinson B, Hall G, Watson RJ, Larocca RA, Abbink P, Peron JPS, De PMAZ, Iampietro MJ, Badamchi-Zadeh A, Maharaj Y, Soliman MES, Hernandez M, Ghersi D, Sanchez R, Tambunan USF, Zahroh H, Utomo BB, Parikesit AA, Niyomrattanakit P, Chen YL, Dong H, Yin Z, Qing M, Glickman JF, Eyer L, Nencka R, Huvarova I, Palus M, Alves M, Gould EA, Moonsamy S, Bhakat S, Soliman MES, Liao C, Sitzmann M, Pugliese A, Nicklaus MC, Thompson JD, Gibson TJ, Higgins DG, Eswar N, Webb B, Marti-Renom MA, Madhusudhan MS, Eramian D, Shen MY, Somarowthu S, Yang H, Hildebrand DGC, Ondrechen MJ, Huang B, Irwin JJ, Shoichet BK, Koes DR, Camacho CJ, Trott O, Olson AJ, Case DA, Cheatham TE, Darden T, Gohlke H, Luo R, Merz KM. Zika virus drug targets: a missing link in drug design and discovery—A route map to fill the gap. RSC Advances. 2016;**6**:68719-68731. DOI: 10.1039/C6RA12142J

[118] Chakraborty S. MEPPitope: Spatial, electrostatic and secondary structure perturbations in the post-fusion Dengue virus envelope protein highlights known epitopes and conserved residues in the Zika virus. F1000Research. 2016;**5**:1150. DOI: 10.12688/f1000research.8853.1

[119] Li C, Deng YQ, Wang S, Ma F, Aliyari R, Huang XY, Zhang NN, Watanabe M, Dong HL, Liu P, Li XF, Ye Q, Tian M, Hong S, Fan J, Zhao H, Li L, Vishlaghi N, Buth JE, Au C, Liu Y, Lu N, Du P, Qin FXF, Zhang B, Gong D, Dai X, Sun R, Novitch BG, Xu Z, Qin CF, Cheng G. 25-hydroxycholesterol protects host against Zika virus infection and its associated microcephaly in a mouse model. Immunity. 2017;**46**:446-456. DOI: 10.1016/j.immuni.2017.02.012

[120] Zhu Z, Gorman MJ, McKenzie LD, Chai JN, Hubert CG, Prager BC, Fernandez E, Richner JM, Zhang R, Shan C, Wang X, Shi P-Y, Diamond MS, Rich JN, Chheda MG. Zika virus has oncolytic activity against glioblastoma stem cells. Journal of Experimental Medicine. 2017. DOI: 10.1084/jem.20171093

Section 2

Epidemiological and Entomological Aspects

Genetic and Geo-Epidemiological Analysis of the Zika Virus Pandemic; Learning Lessons from the Recent Ebola Outbreak

Dimitrios Vlachakis, Louis Papageorgiou and Vasileios Megalooikonomou

Additional information is available at the end of the chapter

http://dx.doi.org/10.5772/intechopen.71505

Abstract

"Outbreak" is a term referring to a virus or a parasite that is transmitted very aggressively and therefore could potentially cause fatalities, as the recent Ebola and Zika epidemics did. Nevertheless, looking back through history, quite a few outbreaks have been reported, which turned out so deadly that essentially changed, molded and literally re-shaped the society as it is today. In the present chapter, differences and similarities between the two most recent outbreaks have been studied, in order to pinpoint and design a trace model that will allow us to draw some conclusions for the connection of those two epidemics. Due to the high dimensionality of the problem, modern and state of the art geo-epidemiological methods have been used in an effort to provide the means necessary to establish the abovementioned model. It is only through geo-epidemiological analysis that it is possible to analyze concurrently a multitude of variables, such as genetic, environmental, behavioral, socioeconomic and a series of related infection risk factors.

Keywords: genetics, evolution, epidemiology, Zika, Ebola

1. Introduction

The *Flaviviridae* is a diverse viral family of more than 100 known viral species, which has a worldwide distribution with several different viral members across the continents of our planet [1]. The first *Flaviviridae* virus outbreak was recorded in 1878 with the yellow fever virus at the city of Memphis, Tennessee, a hometown of 45,000–50,000 people that suddenly

became a ghost town [2]. At the same time, germ theory was still a state of the art concept. Researchers had no idea that the cause of yellow fever was a virus 1000 times smaller in size the human eye could detect, and they definitely had no idea that arthropod vectors were the carriers, and for this case mosquitoes in particular. Yellow fever virus was isolated for the first time in 1927, and it was not until 10 years later that an effective vaccine was developed against the fatal virus [3]. Due to the fact that *Flaviviridae* is a highly mutagenic family of viruses, in subsequent outbreaks, several new members have been identified and reported to share high similarity in structural conservation and molecular characteristics. Overall, the genus *Flaviviridae* comprises four main genera *Hepacivirus*, *Pegivirus*, *Pestivirus* and *Flavivirus* [4, 5].

The genus *Hepacivirus* is the smallest and contains the Hepatitis C Virus (HCV), one of the most fatal human pathogens in the *Flaviviridae* family [6, 7]. The genus *Pegivirus* was recently classified and includes virus species that infect mammals [4, 8]. The genus *Pestivirus* has not been classified as a zoonotic disease yet and contains the bovine viral diarrhea virus (BVD) and the classical swine fever virus (CFSV) [4, 9]. Nevertheless their impact on livestock is strongly connected with the economic and social well-being of many countries. Last but not least the genus *Flavivirus* is the largest one and contains 70 identified human and animal viruses [4] including Dengue virus (DENV), West Nile virus (WNV), Japanese encephalitis virus (JEV), Tick-borne encephalitis virus (TBEV), Yellow fever virus (YFV) and Zika virus (ZIKV) the last of which recently led to the World Health Organization (WHO) in order to declare a global public-health emergency.

Among other common characteristics, the *Flaviviridae* family viruses spread quickly and easily in a tropical environment and attribute that the researchers are now re-examining the current Zika virus outbreak [10, 11]. In fact, this is down to the way viruses pass from one species to another. The viruses within this family fall within significant medical concern as variety of diseases inflict to both humans and animals. The majority of these viruses spread by arthropod vectors such as ticks and mosquitoes and in many cases they can also be transmitted from animals to humans. *Flaviviridae* family viruses can survive for long periods in their hosts by replicating without damaging the host cell. Generally, humans are infected with these viruses by direct contact with infected blood. In America, people have been mostly infected with Zika virus through the bites of the *Aedes aegypti* mosquito [12].

The ability of the *Flaviviridae* family viruses to spread so rapidly is also due to the structure of its viral genome [4, 13]. The *Flaviviridae* are a family of positive, single-stranded enveloped RNA viruses. Their genetic material in the form of ssRNA contains all the information needed in order to make copies and it also mutates more easily [14]. Mistakes in replication can happen in both DNA and RNA, but RNA has fewer systems in place to proofread and correct mutations that may naturally arise [15]. In fact the RNA viruses do not have these correcting mechanisms at all, so the mutations remain and get passed to the next generation [14]. RNA replication has an error rate roughly 10,000 times higher than that of DNA, which means that *Flaviviridae* family viruses evolve much quicker.

Flaviviridae family genera present similarities in the organization of its genetic material, the estimate life cycle, replication and morphology of the viral particles [5]. Virions of *Flaviviridae* family viruses are spherical almost 40–60 nm in diameter. Each virion consists of a lipid envelope which is composed of two or three virus-encoded membrane proteins, a membrane and a small capsid. The genetic material is located within the capsid. The genome includes a single-stranded positive sense RNA molecule of approximately 9.5–12.5 kb. It contains a long open reading frame (ORF) which is found between untranslated regions (UTRs) at 5′ and 3′ ends. All the family members lack 3′ terminal polyadenylated tail. The whole genome is translated into a polyprotein, which is processed post-translationally by host and viral proteases. This polyprotein consists of minimum 10 different products, depending on the classification of the virus that can be divided based on structural and non-structural (NS) proteins [7].

2. General epidemiology

Despite the fact that the Zika virus epidemic in the Americas and the Caribbean is showing signs of a significant slow-down, Zika virus transmission is continued worldwide. To date, 148 areas worldwide related with Zika virus from which 61 areas worldwide have confirmed the transmission of Zika virus disease, 18 areas have evidence of Zika virus circulation and 69 have the potential for Zika virus future transmission [16, 17]. Moreover five countries have reported sexually transmitted Zika cases [18, 19]. There are critical knowledge gaps around Zika virus and a lack of historical reports on its vectors, transmission patterns and geographical distribution. Regardless these challenges, there is a need to better understand the epidemiology of Zika virus transmission in a given area, at a given time in order to allow an assessment of the possibility of Zika virus infection for a number of populations, and to adapt public health recommendations for residents and travelers [17].

The first isolation of Zika virus was detected in 1948 from the mosquito species *Aedes africanus* [20, 21]. Since then, there have been several new cases of Zika virus vectors reported to literature. Collectively, the most frequent mosquito vector is the genus *Aedes* including, *A. aegypti*, *Aedes furcifer*, *Aedes luteocephalus*, *Aedes vitattus* and *Aedes apicoargenteus* [22–24]. The main vector of the Zika virus in South and South-East Asia is considered to be the genus *A. aegypti* of mosquitoes, which also play a major role as the major transmission vector for Dengue virus, Yellow fever virus, Chikungunya virus and many other mosquito-borne *flaviviruses* [25]. Based on reports during the two major outbreaks in the Pacific Islands, new potential vectors of Zika virus could be *Aedes hensilli* and *Aedes polynesiensis* mosquitoes [26, 27]. Moreover, recently another member of mosquitoes within the *Aedes* genus was found to be infected, the *Aedes albopictu* [12, 28]. Zika virus replicates in mosquitoes within 7–10 days and afterwards spreads with high levels of transmission through the salivary glands. However, the superiority of either vector over others in its capacity to optimally transmit Zika virus is yet to be investigated. Despite the Zika virus reservoir is unclear, some researcher believe that Zika virus might be maintained in nature by a sylvatic cycle involving non-human primates, or in a broad members of *Aedes* mosquito's species [23, 24]. Zika virus transmission in urban areas takes place by the anthropophilic mosquitoes *A. aegypti* and *A. albopictus*, from which the *A. albopictus* is particularly worrisome from the *A. aegypti* because of its daily feeding and their custom to bite several hosts during the development cycle of their eggs which makes them very effective as a transmission vectors [21, 29, 30]. Nevertheless, serological studies have established the presence of particular antiviral antibodies in several mammals including elephants, felines and rodents, which confirm that other reservoirs may be involved in the transmission cycle of the Zika virus [21, 31]. Therefore, there is an urgent need to identify vectors and possible vectors in vulnerable area in order to control Zika virus outbreaks.

3. Genetic and geo-epidemiological analysis of the Zika virus

Zika virus has been around since the first human case was diagnosed in 1952 [20]. It only caused sporadic human infections until the first humongous outbreak on the Micronesian island of Yap in 2007 [32]. Even then, the symptoms were reported to share many common characteristics with a mild flu symptoms including fever, rash, headaches and joint and muscle pain. However, during the Zika virus outbreak in Tahiti in 2013 and 2014, a small number of people began to show unusual symptoms that resembled another syndrome called "Guillain-Barré," a condition that causes the immune system to attack a person's nerves, resulting in muscle weakness, tingling and even paralysis [33]. Furthermore, pregnant women infected with Zika virus have also given birth to babies with microcephaly, or smaller than normal heads [34]. It is possible that these new symptoms could be a result of the Zika virus evolution [35]. Zika virus was relatively infrequent and it did not cause a lot of concern, but when new cases were identified with greater frequency and people started to see babies born with microcephaly and nervous system disorders, the public health importance of this virus had to change completely [36].

The sudden appearance and the rapid global spread of the Zika virus have caused confusion regarding the meaning of the term "pandemic" [37]. There is also confusion in the recognition of pandemics when they occur. Some believe that the explosive contagiousness is sufficient to classify a disease as a pandemic. Another opinion claims that the severity of the infection and mortality must be considered. A pandemic is an epidemic of infectious disease that has spread throughout human population across a large region such as multiple continents or even worldwide [38]. The pandemic is not related to the endemic, which is a disease that can be controlled as far as transmissibility is concerned. On the other hand, the pandemic is characterized as a lethal threat to humans, since mortality rates may overcome these wars and accidents. The main feature of the pandemic is that it is communicable.

A look back in history reveals that pandemics are so expansive and deadly that they essentially changed the course of history such as the Black Death or as well known as "Bubonic Plague" in 1339–1351, which affects both Europe and Asia [39]. Painful, egg-sized swellings on the body signaled the infected had only a week to live [40]. Being close to a sick person was enough to get infected. The total number of worldwide deaths is estimated at 75 million people and even today the WHO reports 1000–3000 cases of plague every year [41]. Furthermore, Cholera boomed during 1817–1823 in Asia, Middle East and Africa. The cholera bacterium lurks in contaminated water and infects the small intestine, causing severe diarrhea. About 40% of victims die within hours, due to dehydration. Cholera has killed at least 10 million people. An estimated 2.86 million cholera cases occur annually in endemic countries. Among these cases, there are an estimated 95,000 deaths [42]. Nowadays, HIV/AIDS is still expanded worldwide since 1970, despite the fact that it is originated in central Africa [43, 44]. HIV virus viral genome was identified in the early 1980s, after an epidemic expand among homosexual men in western countries. HIV is transmitted through bodily fluids and blood, and destroys the immune system [45]. HIV infected people with zero immune response are easier to die of other diseases such as cancer or pneumonia, or from simple infections. Since there is no cure, anti-retroviral drugs are granted to the patients to prevent the virus from spreading and killing more immune cells. HIV and generally AIDS have killed more than half a million North Americans [46]. In Southern Africa, it is estimated that 25 million have lost their lives, with 35 million more still infected, including many children born with the virus [47].

The World Health Organization (WHO) has produced a six-stage (phasing) classification that describes the process by which a novel influenza virus moves from the first few infections in humans through a pandemic [48]. The stages do not relate to how sick the person gets or how many people are infected by the virus. Instead, they relate to where it is located and how it is spread from one area to another. Firstly, viruses circulate within animals only. No human infection has resulted from the animal virus. Secondly, if an animal virus infects humans, it means that this virus has mutated, so there is a basic level of pandemic threat. Thirdly, small groups of people in a region are infected by the virus. It is possible to transmit the virus outside the boundaries of the community if others outside that community come into contact with those humans who are infected. The next phase is when the outbreaks of the virus are increasing dramatically in many communities in a short period of time. A stage before the pandemic is when the transmission of the virus from human to human has been recorded in at least two countries. Finally, the pandemic phase means a global pandemic is underway. Illness is widespread and governments and worldwide organizations are actively working to curtail the spread of the disease, and to help the population deal with it, by using preventive measures [49].

Due to the nature of pandemics, we will never be fully protected despite any development in antivirals and viral medicine [50]. The following categories are the most ominous: Viral hemorrhagic fevers, including Ebola and Marburg Virus, have the potential to turn into pandemics. Influenza—the recent discovery of the H5N1 (Avian Flu) is an example of this. The strain was spotted in Vietnam in 2004; the ability of the virus to potentially combine with human flu viruses is a concern to scientists. Ebola—the largest Ebola epidemic the world has ever seen is still ongoing. Huge efforts have been made in order to prevent it from turning into a pandemic. Viral diseases occupy a dominant position in pandemics. The majority of deaths worldwide caused by viral pandemics, the mortality of diseases that belong to *Flaviviridae* family are shown in **Table 1**. A major problem is the immediate and prompt response for protection before a disease evolves into a pandemic. The progress of science in drug discovery is rapid, scientists replace the traditional methods of in vivo/in vitro trial error testing and focus more on techniques of rational drug design based on the structure which is efficient, fast and of lower cost [51].

To date there is neither a drug nor a specific vaccination available against Zika virus, Ebola or HIV virus. Thus there is a great need for the development of novel antiviral strategies.

Virus	Case fatality rate (CFR)	Diagnosed (per year)	Deaths (per year)
Yellow Fever (YFV)	7.5%	200.000	30.000
Dengue Fever (DENV)	40–50% (without treatment) 1–5% (with treatment)	50–528 million	20.000
Japanese encephalitis virus (JEV)	0.3–60%	70.000	20.400
Tick-borne encephalitis virus (TBE)	20–40% far Easter 2–3% E.U, U.S.A	10.000–15.000	1000
Hepatitis C virus (HCV)	9%	3–4 million	350.000
West Nile virus (WNV)	3–15%	3.000	100
Zika virus (ZIK)	—	1,5 million	18 (+71*)
	*cases of deaths with microcephaly and/or central nervous system (CNS) malformation in newborns.		

Table 1. Case fatality rate (CFR) of viruses in *Flaviviridae* family. Microcephaly and other fetal malformations potentially associated with Zika virus infection or suggestive of congenital infection, have been reported in seven countries or territories. Particularly, from October 2015 until January 2016, 4783 cases of microcephaly have been reported in Brazil. On 29 April the first American died of complications related to the Zika virus, health officials of CDC reported.

Viruses have a relatively simple structure. They contain nucleic acids in a capsid. The mechanism used for their transmission is based on the host cell. In the end, the host cell is destroyed after the creation of multiple copies of the virus. An antiviral strategy for *Flaviviridae* family is to identify which proteins/enzymes are involved in viral replication. Important enzymes are the viral helicase and viral polymerases, NS3 and NS5, respectively [52]. The knowledge that the helicase plays a key role in the translation of Zika virus leads to the establishment and the design of homologous helicase models in order to be used for the rational design of an anti-Zika pharmacophore [53]. Also it has been shown that IFN, ribavirin, 6-azauridine and glycyrrhizin have the ability to inhibit infection of VERO cells induced with IFN to be more effective [54]. Moreover, the combination of IFN and ribavirin may be more effective in the *Flavivirus* genus. Researchers have shown that propoxy derivatives are good candidates for drugs against HCV [55]. Anti-malaria hydroxychloroquine indicated inhibition of dengue virus infection and has been safely used during pregnancy as well as amodiaquine and tetracyclines [56].

Ebola virus disease (EVD) is a non-segmented, negative-sense, single-stranded RNA virus that resembles rhabdoviruses and paramyxoviruses in its genome organization and replication mechanisms. It is a member of the family Filoviridae, based upon their filamentous structure. The genus Ebola virus is divided into five species (Zaire, Sudan, Ivory Coast, Bundibugyo and Reston). Among them, the first four cause disease in humans. The disease has a high risk of death, killing between 25 and 90% of those infected, with an average of about 50% [57]. The disease was first identified in 1976 in two simultaneous outbreaks, one in Nzara and the other in Yambuku, a village near the Ebola River from which the disease took its name. The old virus differs from today by 3%. Ebola outbreaks frequently make their appearance in sub-Saharan Africa. So far the greatest epidemic ever recorded was in western Africa (December 2013–January 2016) with 11,315 confirmed deaths. Symptoms of Ebola typically include fever, severe headache, muscle pain, weakness and sore throat. Late-stage symptoms of Ebola virus may include vomiting, diarrhea, redness in the eyes, swelling of the genitals, internal and external bleeding. Typically, symptoms appear 8–10 days after exposure to the virus, but the incubation period can span from 2 to 21 days. Recovery may begin between 7 and 14 days after first symptoms. Death, if it occurs, follows typically 6–16 days from first symptoms and is often due to low blood pressure from fluid loss [58].

Ebola virus is transmitted from human to human by close contact with infected patients and virus-containing body fluids. Specifically, it spreads through direct contact with blood or body fluids (including but not limited to feces, saliva, sweat, urine, vomit and semen) of a person who is sick with or has died from Ebola [59]. Ebola can also be spread through needle sticks and contact with objects (like needles and syringes) that have been contaminated with the virus. According to WHO, you can also contract the virus by handling a sick or dead wild animal that has been infected with it. In the countries of West Africa, transmission through infected animals has been observed, usually infecting bats or wild animals as prey. The Ebola virus is also a sexually transmitted disease, transmitted by semen (oral, vaginal and anal sex) [60]. High-risk groups, except the population of sub-Saharan Africa are people surrounded by patients either in the family or in the professional environment. Scientists observing the virus have not seen any evidence to suggest that the Ebola virus may be mutating to become more contagious.

Ebola is difficult to be diagnosed when a person is first infected because the early symptoms, such as fever, are also symptoms of other diseases. The main question doctors consider is whether the person has been in one of the countries in West Africa within the last 21 days, then tests of blood and tissues can diagnose Ebola. Laboratory diagnosis of Ebola is achieved in two ways: detection of infectious particles (or particle components) in affected individuals and measurement of specific immune responses to Ebola virus [61]. To date, no FDA-approved vaccine or medication is available against Ebola virus. Albeit, when some basic interventions are used early, they can improve the chances of survival. The basic interventions used are providing intravenous fluids (IV) and balancing electrolytes (body salts), maintaining oxygen status and blood pressure and treating other infections if they occur, medications to treat shock and pain medications.

Since no therapy has been approved, an important issue that occurs is prevention. Preventive methods are vital. WHO and other global health organizations have suggested several types of protection, some of them are wash your hands with soap and water or an alcohol-based hand sanitizer and avoid contact with blood and body fluids, isolate the patient, wear protective clothing, dispose of needles and syringes safely, use safe burial practices and avoid facilities in West Africa where Ebola patients are being treated.

To date, the largest outbreak of the Ebola virus was in West Africa. In March 2014, Guinea, Liberia and Sierra Leone were the countries where the major outbreak of Ebola appeared. In early October of the same year, the first transmission of the virus to occur outside Africa was reported, later other patients were identified in Europe and America. The numbers show the extent of the problem. In the period 2013–2015 in Liberia, Sierra Leone, Guinea, Nigeria, Mali, United States, Senegal, Spain, United Kingdom and Italy, there were 11,315 deaths with case fatality rate of up to70–71%. The United National health agency in December 29, 2015 declared the end of the Ebola virus transmission in Guinea, Liberia and Sierra Leone, where the epidemic began killing 13,000 people [62].

Zika virus was originally isolated in 1947 from the blood of a Rhesus monkey for yellow fever research conducted in the Zika forest, Uganda, from which the disease took its name. Then the virus was recovered again from humans and mosquitoes in the countries of Central Africa and Malaysia. The Zika virus belongs to the genus *Flavivirus* of the *Flaviviridae* family. It is a single-stranded RNA virus, with a shell and the shape of a sphere. Genomic comparisons have revealed that it has two major lineages: Asian and African on the basis of their nucleotide sequences. Typical symptoms are fever, headache, maculopapular rash that starts on the face and spreads to the whole body, redness in the eyes or conjunctivitis. Symptoms are generally mild and start about 2 or 7 days after the infection. Furthermore in some cases, lack of fever and often fewer symptoms such as muscle pain and arthralgia have been observed.

The Zika virus is a vector-borne disease, transmitted by several *Aedes* (stegomyia) species of mosquitoes, having been isolated in *A. albopictus* mosquitoes and *A. aegypti*. Other reported modes of transmission are through mother to fetus, through laboratory exposure, blood transfusion and through sexual contact. Transmission through transplantation is possible but not proven. The Zika virus has been detected in human blood, saliva, semen and urine. It has been confirmed that it remains after 62 days of infection in the patient's semen and urine. A possible mode of transmission is considered to be the bite of infected animals with the virus. All these transmission modes of Zika virus make it a challenge for science and for the whole world.

Since the virus can be detected in human body fluids, various methods have been developed for diagnosing it, through saliva or urine samples that have been collected over 3–5 days after the onset of symptoms. It is possible to isolate the viral genome and detect the nucleic acid by reverse transcriptase-polymerase chain reaction. Serological tests like ELISA or immunofluorescence are also widely used. CDC had developed an ELISA technique to detect specific anti-Zika IGM during the epidemic in Yap in 2007 [63].

Currently, no specific treatment or vaccination is available for Zika virus. Treatment of symptoms may include rest, fluids, antipyretics (the CDC advice against using aspirin or other nonsteroidal anti-inflammatory medication) and analgesics. In addition, the CDC has advised that pregnant women who are diagnosed with Zika virus should be considered for the monitoring of fetal growth and anatomy every 3–4 weeks, because of Zika virus correlation with infant microcephaly. Prevent Zika virus by avoiding mosquito bites. Mosquitoes that spread Zika virus bite mostly during the daytime. Additionally, sexual transmission of Zika virus can be prevented by using condoms or not having sex.

The geographical spread of the virus Zika virus beyond Africa and Asia was reported in 2007. The first major expansion with 185 cases occurred in 2007 on the Yap Islands of the Federated States of Micronesia. Scientists assume that the introduction of the virus in Yap probably came from an infected mosquito or human from Southeast Asia. In 2013, the virus broke out again near the French Polynesia. In May 2015, the Ministry of Health of Brazil sounded the alarm because 14 states of the country were affected by the virus. Scientists suspect that Zika virus is the cause of 2400 cases of microcephaly and 29 infant deaths in Brazil only in 2015. Currently confirmed cases are in 31 countries of South America, 5 countries in Oceania/Pacific Islands and in Care Verde. **Figure 1** represents all counties with active Zika virus this period March 2016. The number is expected to increase due to the travel movements (holidays, Olympic Games, pilgrimage to Mecca). Furthermore, 80% of cases will not be diagnosed because they do not have any symptoms. The distribution of *A. aegypti* and *A. albopictus* has remarkable parallelisms with the spread of Zika virus.

The climate change has a great affect on the transmission of the Zika virus. *A. aegypti* is a known vector of several viruses, including dengue fever, yellow fever and Zika virus. They have identified hundreds of cases in Europe, some ended in death. Probably, it comes from Africa, having been transferred globally through trade and sailing ships. Now found in tropical, subtropical even in temperate parts. It is relatively small, with black and white patterns. A distinctive external feature is the presence of a silver lyre-shaped sign on the chest. *A. aegypti* has four stages during its life: egg, larva, nymph and adult. In the first three stages, it lives in the water and in the last in the air. Females are hematophagous. They lay their eggs in temperatures between 25 and 29°C in small water-filled containers. In order to survive in a region, they must have adequate temperature and water availability. Humidity is vital not only for the eggs but also for the adults. They prefer to host mammals ideally humans [64] and artificial ponds. *A. aegypti* populations exist in northern Brazil, Southeast Asia, India, Greece, Spain and temperate regions of North America. Another type of mosquito that transmits diseases and belongs to the same family is *A. albopictus*. It originates from the tropical forests of South-East Asia. They have the same appearance with *A. aegypti*, and the diagnostic feature is the presence of a median silver-scale line against a black background on the thorax. Their

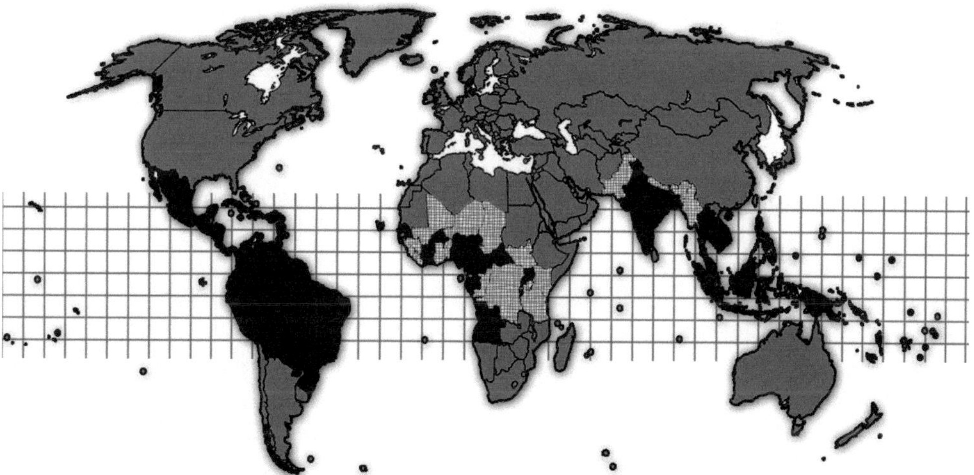

Figure 1. Areas with active Zika virus (colored dark grey, Category 1), areas with evidence of Zika virus (marked with small squares , Category 2) and areas with potential of Zika virus transmission (marked with large squares, Category 3). Figure was constructed based on WHO report of March 2017 about Zika virus.

eggs can survive at −10°C, making them more resistant compared to *A. aegypti*. *A. albopictus* populations expand geographically over the years. The main reason for the enhanced distribution of vectors *A. aegypti* and *A. albopictus* is climate change. Climate change has already changed the distribution of certain bodies of animals and is expected to influence it even further. According to the European Environment Agency, the global average temperature has increased by 0.74°C and the water level is increasing by 1.8 mm every year due to the fact that the ice in the Arctic melts at 2.7% each decade. Climate change may cause changes to the period and intensity of infectious diseases. The rise in temperature allows the survival of insects at higher altitudes. Proof of this can be malaria, which has expanded into areas, where previously it did not succeed. Dengue fever has also been seen in Puerto Rico, Florida, Gulf Coast states and Hawaii, places that had not usually been affected. Rising temperatures in the southern coast of the United States of America areas in Florida, Hawaii and the Coast state the following potential residence of *A. aegypti* together with Zika virus. The wind can reduce mosquito bites but also extend their flights. The increased rainfall creates small natural ponds, excellent conditions for reproduction and survival of mosquitoes. All these events create a clear picture that the protection of global health comes from environmental protection.

Another major factor that has an immediate effect on the transmission of the viruses is the mass population transfer from less safe areas to unaffected areas where they could find breeding ground for the expansion of the viruses [65]. The communities and the international organizations should not be negative toward the population transfer from poorer to richer countries or from countries in war to countries in peace as long as extra measures are being taken like a database for all immigrants that enter another country, On the other hand, the technological achievements of the twenty-first century have annihilated the distances and many travelers can be easily transferred all over the globe [66]. Thus, new measures are necessary for the population control in all the vulnerable continents like Europe and Asia. In addition, an online webpage should be created in order to update on a daily basis about the new crusts in each continent for every virus from the ones mentioned above. A new more detailed profile for any incoming stigma should be created for every crust. The study of every specimen for the immediate tracking of dangerous mutations of the virus and the immediate report of different symptoms from the ones expected in case of detecting a familiar virus. Phylogeography of the viruses is a new revolutionary idea through which we can comprehend and compare the nature of the viruses for the better treatment in case of random outbreaks [67].

What do Zika virus and Ebola have in common? Although both Ebola and Zika virus have been known for decades, in the last period they have been in the forefront of international attention not only for their huge spread but also for the serious consequences they induced. Ebola is characterized as a deadly and highly contagious virus. On the other hand, Zika virus is not fatal but scientists associate it with microcephaly infants and Guillain-Barré syndrome. According to the World Health Organization, both Ebola and Zika virus are infectious diseases that began and spread from animals to humans. Ebola is transmitted through infected animal and body fluids. Zika virus mainly through mosquito bites, from mother to fetus and isolated cases have reported transmission through semen and blood. The Ebola virus can only be transmitted when symptoms are present, the virus incubation period is from 2 to

21 days. We do not know if Zika virus is transmitted asymptotically, the incubation period of Zika virus has not been calculated exactly, but probably a few days to a week. Both viruses have mutated. Ebola causes several symptoms such as fever, vomiting and excessive bleeding while Zika virus exhibits mild symptoms or none, since 80% of the cases show no symptoms. It is worth to mention that there is neither treatment nor vaccine for both Ebola and Zika virus. Ebola is characterized by high mortality rates ranging from 50 to 90%. There are similarities in how these two epidemics unfold. Both of them were reported much later, many deaths from Ebola until declared an emergency for public health, and many microcephaly cases which are believed to be associated with Zika virus. Interesting is the similarities that geographical areas have which were inflicted by the two viruses. Ebola started and was mainly reported in Guinea, Sierra Leone and Liberia. These countries have a shortage of drinking water, residents have a poor diet, living in unsightly and unsanitary conditions and made use of pesticides and chemicals. Most cases of Zika virus are found in Brazil, Columbia and El Salvador, poor countries with large rural areas, residents have a poor diet with a lack of vitamin A. In Brazil, the use of pesticides has increased and has forbidden chemicals from other countries. Similarities and differences between Ebola and Zika virus are shown in **Table 2**.

Similarities	Differences
Both of them have been known by scientists for decades.	Zika virus typically leads to mild febrile cases, with most being asymptomatic, Ebola has severe symptoms.
They are an animal origin in other words, a zoonoses.	Ebola has high mortality rate Zika virus not.
No treatment or vaccine	The incubation period of Ebola 2–21 of Zika virus few days up to a week
They spread in poor countries	Zika virus cannot be transmitted through "casual contact", as ebola
Had a rapid spread	The Ebola can only be transmitted when symptoms are present.
They became epidemics in recent years.	
Both viruses have mutated.	

Table 2. Similarities and differences of viruses Zika virus end Ebola.

4. Conclusion

What Ebola and Zika virus should teach us is that we cannot assume that pathogenic viruses will continue to behave the same way without being mutated. Both viruses have similarities with how they spread. However, we should be mindful that there are some differences between Zika virus and Ebola based on fatality and modes of transmission. Zika virus like many of its cousins (WNV, dengue and chikungunya) will continue to exist and threaten mankind, until answers are provided to several open questions: Are there modes of transmission other than through the vector? Are mosquito species other than Aedes is involved in the urban cycle? Can person-to-person contamination occur through saliva? Can congenital or sexual transmission occur? What is the rate of transmission by blood transfusions? Is ZIKV capable of establishing a chronic infection? Is there the generation of a long-lived protective immune response? Is there the possibility of re-infection? These questions must be urgently answered to allow the effective design of strategies to prevent and/or treat ZIKV transmission and infection and will demand a collective and coordinate basic research initiative to address these issues.

Author details

Dimitrios Vlachakis[1,2]*, Louis Papageorgiou[3] and Vasileios Megalooikonomou[1]

*Address all correspondence to: dvlachakis@bioacademy.gr

1 Computer Engineering and Informatics Department, University of Patras, Patras, Greece

2 Genetics Lab, Department of Biotechnology, Agricultural University of Athens, Athens, Greece

3 Department of Informatics and Telecommunications, National and Kapodistrian University of Athens, University Campus, Athens, Greece

References

[1] Calzolari M et al. Insect-specific flaviviruses, a worldwide widespread group of viruses only detected in insects. Infection, Genetics and Evolution. 2016;**40**:381-388

[2] Wright FM. The 1878 yellow fever epidemic in Memphis. Journal of the Mississippi State Medical Association. 2001;**42**(1):9-13

[3] Koide SS. Hideyo Noguchi's last stand: The yellow fever Commission in Accra, Africa (1927-8). Journal of Medical Biography. 2000;**8**(2):97-101

[4] Papageorgiou L et al. An updated evolutionary study of Flaviviridae NS3 helicase and NS5 RNA-dependent RNA polymerase reveals novel invariable motifs as potential pharmacological targets. Molecular BioSystems. 2016;**12**(7):2080-2093

[5] Papageorgiou L et al. Structural models for the design of novel antiviral agents against Greek goat encephalitis. PeerJ. 2014;**2** e664

[6] Blackard JT et al. Acute hepatitis C virus infection: A chronic problem. Hepatology. 2008;**47**(1):321-331

[7] Vlachakis D, Koumandou VL, Kossida S. A holistic evolutionary and structural study of flaviviridae provides insights into the function and inhibition of HCV helicase. PeerJ. 2013;**1**:e74

[8] Da Mota LD et al. Prevalence of human pegivirus (HPgV) infection in patients carrying HIV-1C or non-C in southern Brazil. Journal of Medical Virology. 2016;**88**(12):2106-2114

[9] Moennig V, Becher P. Pestivirus control programs: How far have we come and where are we going? Animal Health Research Reviews. 2015;**16**(1):83-87

[10] Campos GS, Bandeira AC, Sardi SI. Zika Virus Outbreak, Bahia, Brazil. Emerging Infectious Diseases. 2015;**21**(10):1885-1886

[11] WHO, Zika virus outbreaks in the Americas. Wkly Epidemiol Rec. 2015;**90**(45):609-610

[12] Chouin-Carneiro T et al. Differential susceptibilities of *Aedes Aegypti* and *Aedes albopictus* from the Americas to Zika virus. PLoS Neglected Tropical Diseases. 2016;**10**(3) e0004543

[13] Thurner C et al. Conserved RNA secondary structures in Flaviviridae genomes. The Journal of General Virology. 2004;**85**(Pt 5):1113-1124

[14] Elena SF, Sanjuan R. Adaptive value of high mutation rates of RNA viruses: Separating causes from consequences. Journal of Virology. 2005;**79**(18):11555-11558

[15] Drake JW, Holland JJ. Mutation rates among RNA viruses. Proceedings of the National Academy of Sciences of the United States of America. 1999;**96**(24):13910-13913

[16] WHO, Zika Virus; Microcephaly; Guillain-Barre Syndrome. WHO Report, 2017. WHO/ZIKV/SUR/17.1

[17] WHO, Zika Virus Country Classification Scheme. WHO Interim guidance 2017. WHO/ZIKV/SUR/17.1

[18] Musso D et al. Potential sexual transmission of Zika virus. Emerging Infectious Diseases. 2015;**21**(2):359-361

[19] D'Ortenzio E et al. Evidence of sexual transmission of Zika virus. The New England Journal of Medicine. 2016;**374**(22):2195-2198

[20] Dick GW, Kitchen SF, Haddow AJ. Zika virus. I. Isolations and serological specificity. Transactions of the Royal Society of Tropical Medicine and Hygiene. 1952;**46**(5):509-520

[21] Haddow AD et al. Genetic characterization of Zika virus strains: Geographic expansion of the Asian lineage. PLoS Neglected Tropical Diseases. 2012;**6**(2):e1477

[22] Li MI et al. Oral susceptibility of Singapore Aedes (Stegomyia) aegypti (Linnaeus) to Zika virus. PLoS Neglected Tropical Diseases. 2012;**6**(8):e1792

[23] Grard G et al. Genomics and evolution of Aedes-borne flaviviruses. The Journal of General Virology. 2010;**91**(Pt 1):87-94

[24] Wolfe ND et al. Sylvatic transmission of arboviruses among Bornean orangutans. The American Journal of Tropical Medicine and Hygiene. 2001;**64**(5-6):310-316

[25] Powell JR, Tabachnick WJ. History of domestication and spread of *Aedes aegypti*--a review. Memórias do Instituto Oswaldo Cruz. 2013;**108**(Suppl 1):11-17

[26] Musso D. Zika virus transmission from French Polynesia to Brazil. Emerging Infectious Diseases. 2015;**21**(10):1887

[27] Craig AT et al. Acute flaccid paralysis incidence and Zika virus surveillance, Pacific Islands. Bulletin of the World Health Organization. 2017;**95**(1):69-75

[28] Heitmann A et al. Experimental transmission of Zika virus by mosquitoes from central Europe. Euro Surveillance. 2017;**22**(2)

[29] Delatte H et al. Blood-feeding behavior of *Aedes albopictus*, a vector of Chikungunya on la Reunion. Vector Borne and Zoonotic Diseases. 2010;**10**(3):249-258

[30] Amraoui F, Vazeille M, Failloux AB. French *Aedes albopictus* are able to transmit yellow fever virus. Euro Surveillance. 2016;**21**(39)

[31] Darwish MA et al. A sero-epidemiological survey for Bunyaviridae and certain other arboviruses in Pakistan. Transactions of the Royal Society of Tropical Medicine and Hygiene. 1983;**77**(4):446-450

[32] Duffy MR et al. Zika virus outbreak on Yap Island, Federated States of Micronesia. The New England Journal of Medicine. 2009;**360**(24):2536-2543

[33] Waehre T et al. Zika virus infection after travel to Tahiti, December 2013. Emerging Infectious Diseases. 2014;**20**(8):1412-1414

[34] Rubin EJ, Greene MF, Baden LR. Zika virus and microcephaly. The New England Journal of Medicine. 2016;**374**(10):984-985

[35] Konda S, Dayawansa S, Huang JH. The evolution and challenge of the Zika virus and its uncharted territory in the neurological realm. Journal of Neuroinfectious Diseases. 2016;**7**(2)

[36] Faye O et al. Molecular evolution of Zika virus during its emergence in the 20(th) century. PLoS Neglected Tropical Diseases. 2014;**8**(1):e2636

[37] Ribeiro GS, Kitron U. Zika virus pandemic: A human and public health crisis. Revista da Sociedade Brasileira de Medicina Tropical. 2016;**49**(1):1-3

[38] Pratt MK. Pandemics. Essential Issues. Edina, Minn: ABDO Pub; 2011. 112 p

[39] Duncan CJ, Scott S. What caused the black death? Postgraduate Medical Journal. 2005;**81**(955):315-320

[40] Cohn SK Jr. Epidemiology of the black death and successive waves of plague. Medical History. Supplement. 2008;**27**:74-100

[41] Matt J, Keeling CAG. Bubonic plague: A metapopulation model of a zoonosis. Proceedings of the Royal Society B: Biological Sciences. 2000;**267**:2219-2230

[42] Mohammad Ali ARN, Lopez AL, Sack D. Updated global burden of cholera in endemic countries. PLoS Neglected Tropical Diseases. 2015;**9**

[43] Simon V, Ho DD, Abdool Karim Q. HIV/AIDS epidemiology, pathogenesis, prevention, and treatment. Lancet. 2006;**368**(9534):489-504

[44] Worobey M et al. 1970s and 'Patient 0' HIV-1 genomes illuminate early HIV/AIDS history in North America. Nature. 2016;**539**(7627):98-101

[45] Eger KA, Unutmaz D. The innate immune system and HIV pathogenesis. Current HIV/AIDS Reports. 2005;**2**(1):10-15

[46] Pellowski JA et al. A pandemic of the poor: Social disadvantage and the U.S. HIV epidemic. The American Psychologist. 2013;**68**(4):197-209

[47] Jamison DT, World Bank. Disease and Mortality in Sub-Saharan Africa. 2nd ed. Washington, D.C.: World Bank; 2006 xxii, 387 p

[48] Taubenberger JK, Morens DM. The pathology of influenza virus infections. Annual Review of Pathology. 2008;**3**:499-522

[49] WHO. Transcript of virtual press conference with Gregory Hartl, WHO Spokesperson for Epidemic and Pandemic Diseases, and Dr Keiji Fukuda, Assistant Director-General ad Interim for Health Security and Environment, World Health Organization 2009; Available from: http://www.who.int/mediacentre/influenzaAH1N1_presstranscript_20090526.pdf

[50] Vlachakis D, Karozou A, Kossida S. An update on virology and emerging viral epidemics. Journal of Molecular Biochemistry. 2013;**2**:80-84

[51] Papageorgiou L et al. Computer-Aided Drug Design and Biological Evaluation of Novel Anti-Greek Goat Encephalitis Agents. International Journal of Systems Biology and Biomedical Technologies. 2014;**2**:1-16

[52] Loukatou S, Papageorgiou P, Vlachakis D. Optimisation of a potent series of HCV helicase drug candidates. Journal of Molecular Biochemistry. 2015;**4**:1-4

[53] Papageorgiou L et al. Structural models for the design of novel antiviral agents against Greek Goat Encephalitis. Peer J. 2014;**2**:e664

[54] Crance JM et al. Interferon, Ribavirin, 6-Azauridine and Glycyrrhizin: antiviral compounds active against pathogenic flavivirus. Antiviral Research. 2003;**58**(1):73-79

[55] Mazzei M, Nieddu E, Miele M, Balbi A, Ferrone M, Fermeglia M, Mazzei MT, Pricl S, Paolo La Colla, Marongiu F, Cristina Ibbac, Roberta Loddoc. Activity of Mannich bases of 7-hydroxycoumarin against Flaviviridae. Bioorganic & Medicinal Chemistry. 2008;**16**:2591-2605

[56] Wang LF et al. Hydroxychloroquine-inhibited dengue virus is associated with host defense machinery. Journal of Interferon & Cytokine Research. 2015;**35**(3):143-156

[57] Loukatou S et al. Ebola virus epidemic: a deliberate accident? Journal of Molecular Biochemistry. 2014;**3**:72-76

[58] Singh SK, Ruzek D. Viral Hemorrhagic Fevers. CRC Press, Taylor & Francis Group; 2014

[59] Funk DJ, Kumar A. Ebola virus disease: An update for anesthesiologists and intensivists. Canadian Journal of Anesthesia. 2014;**62**(1):80-91

[60] Anna Thorson PF, Lofthouse C, Broutet N. Systematic review of the literature on viral persistence and sexual transmission from recovered Ebola survivors: Evidence and recommendations. BMJ Open. 2016;**6**(1)

[61] Miguel J. Martinez AS, Hurtado JC, Kilgore PE. Ebola Virus Infection: Overview and Update on Prevention and Treatment. 2015. p. 1-26

[62] Akash Mali BS, Pooja J, Dwaraka M, Vaclovas J. Past, present and future about Ebola virus diseases: An updated review. Journal of Pharmacy Practice and Community Medicine. 2016;**2**(2):35-39

[63] Nitin Wahi NN, Sharma S. Zika virus :- a potential threat towards congenital anomalies. 2nd international conference on new challenges in biotechnology and molecular biology in the context of 21st century. At Agra. 2016:2

[64] Saifur RG, Dieng H, Hassan AA, Salmah MR, Satho T, Miake F. Changing domesticity of *Aedes aegypti* in northern peninsular Malaysia: Reproductive consequences and potential epidemiological implications. PLoS One. 2012;**7**(2)

[65] Shackelton LA et al. JC virus evolution and its association with human populations. Journal of Virology. 2006;**80**(20):9928-9933

[66] Parrish CR et al. Cross-species virus transmission and the emergence of new epidemic diseases. Microbiology and Molecular Biology Reviews. 2008;**72**(3):457-470

[67] Holmes EC. Evolutionary history and phylogeography of human viruses. Annual Review of Microbiology. 2008;**62**:307-328

Chapter 4

An Integrated Intervention Model for the Prevention of Zika and Other *Aedes*-Borne Diseases in Women and their Families in Mexico

Norma Pavía-Ruz, Silvina Contreras-Capetillo, Nina Valadéz-González, Josué Villegas-Chim, Rafael Carcaño-Castillo, Guillermo Valencia-Pacheco, Ligia Vera-Gamboa, Fabián Correa-Morales, Josué Herrera-Bojórquez, Hugo Delfín-González, Abdiel Martín-Park, Guadalupe García Escalante, Gonzalo Vázquez-Prokopec, Héctor Gómez-Dantés and Pablo Manrique-Saide

Additional information is available at the end of the chapter

http://dx.doi.org/10.5772/intechopen.71504

Abstract

We describe and discuss the rationale, design and current implementation of a model for an integrated intervention for the primary and secondary prevention of Zika and other *Aedes*-borne diseases and sexually transmitted infections that impact reproductive health, pregnancy and perinatal life stages in women and their families in Merida, Mexico. The intervention includes enhanced surveillance of pregnant women, implementation of communication strategies to promote good practices to prevent disease transmission, determination of the frequency of structural anomalies detected prenatally in the foetus, umbilical cord and placenta of pregnancies diagnosed with ZIK infection, detection of ZIKV and other arboviruses/viruses in products of abortion, in-utero and early newborn, provision of *Aedes aegypti*-proof houses" (protecting homes with door/window screens with insecticide to prevent the access of mosquitoes), mosquito repellents, larvicides and education/promotion of best practices for the prevention of infection with dengue (DENV), chikungunya (CHIKV) and Zika (ZIKV) and an anthropological studies on sociocultural factors of pregnant women associated with ZIKV. This intervention is being developed in a population of 10,000 people of the city of Merida and with the participation of a multidisciplinary group of public health professionals in collaboration with the Ministry of Health and the Government of Yucatan.

Keywords: Zika, dengue, chikungunya, prevention, *Aedes*-borne diseases, *Aedes aegypti*, intervention, house screening, sexual transmission, pregnant women

1. Introduction

Dengue (DEN), chikungunya (CHIK) and now Zika (ZIKV) are *Aedes*-borne diseases (ABD) in different transmission stages in southern Mexico, that is, endemic (DEN), epidemic (CHIK) and emergent (ZIKV), that account for increasing number of cases, chronic disability conditions and deaths along with the threat of increasing numbers of congenital disorders. While ABD face the operational problems associated to the control the mosquito vector *Aedes aegypti*, ZIKV infection has also a non-vectorial transmission route: vertical (transplacental and birth), sexual and/or by blood transfusion which may compromise the health of the mother, fetus and the newborn [1].

The first outbreak of ZIKV outside of Africa and Asia was in Micronesia and affected 72% of the population in 2007 [2] and was confirmed by Centers for Disease Control and Prevention (CDC) [2, 3]. Transmission of (ZIKV) spread to Malaysia, the Philippines and French Polynesia. In 2013–2014, French Polynesia reported a ZIKV outbreak reaching 60% of the popu-lation [2–5]. In 2015, Brazil had an outbreak of ZIKV with 134,057 confirmed cases, were fatal and 80% were asymptomatic [6]. Colombia reported more than 65,000 positive cases of ZIKV during the first half of 2016 [7]. The outbreak of ZIKV in Mexico started at the end of 2015 in Chiapas where 8842 confirmed cases were reported by June of 2017. The Yucatan state, also in south Mexico, has reported 1295 cases and most of confirmed Zika infections in pregnant women (921 confirmed cases) in Mexico [6, 8, 9].

Recent ZIKV epidemiological alerts associate ZIKV/pregnancy with fetal/neonatal morbidity and mortality because there was a marked increase in newborns that developed microcephaly during the ZIKV outbreak in Brazil. By the end of 2015, detection by RNA sequencing of ZIKV from amniotic fluid of foetuses with disorders in the central nervous system was the key for the World Health Organization (WHO) to declare the ZIKV a public health emergency [10–12]. Actually, congenital Zika syndrome (CZS) involves craniofacial malformations (microcephaly, craniofacial anomalies and facial dysmorphism), foetal brain disruption sequence and neural tube defects [13–16]. DEN and CHIK as well, are associated with spontaneous abortion, prematurity and congenital diseases, and the newborn may present thrombocytopenia, hepatomegaly and develop neurological disorders.

The virus has been documented in body fluids as blood, tears, breast milk, vaginal discharge, semen and tissues from abortions and stillbirths [17, 18]. Confirmation of infectious diseases through laboratory diagnostic tests are often necessary; the arrival of ZIKV infection made that the tools developed for other flaviviruses such as DEN, based on immune response (IgM, IgG), would generate false positives to ZIKV. There are several working groups that are developing more specific and sensitive IgM and IgG tests, so it is essential to be definite of the final diagnosis, due to the possible consequences that could involve a false positive or false negative diagnosis. The polymerase chain reaction (PCR) has become the main mechanism for the diagnosis of ZIKV nuclei acid; however, compared to the viral particles of DEN and CHIK, the viral particles of ZIKV are low and therefore the period to give a positive diagnosis in serum is very short and challenging. Other possibilities are detection in urine saliva, breast milk and semen, in pregnant women and their spouses, where a longer viremia is reported [17–20].

This situation challenges vector control and the surveillance system, the prevention of transmission and the organization of health care systems that should effectively deal with the concurrence of three diseases with different clinical outcomes and a common vehicle of transmission. Major challenges include the establishment of a comprehensive communication strategy to sensitize and provide information regarding the risk of ZIKV infection in pregnant women, and the early detection and confirmation of arboviral diseases in the target group (pregnant women) in order to launch the preventive and control interventions. Additional challenges include the improvement or prenatal, pregnancy and newborn care in primary health care centers and hospitals as well as strengthening the surveillance of ABD in these areas.

We need to implement integrated strategies that can improve detection of infection and special medical care for infected women, the follow-up of newborn at risk and the best preventive and vector control measures. Herein, we describe and discuss the design and current implementation of an integrated model for primary and secondary preventive interventions for ZIKV, along with other ABD and sexually transmitted infections, that impact reproductive health, pregnancy and perinatal life stages in women and their upspring in Merida, Mexico.

We describe the rationale, components, implementation and evaluation of this intervention, which includes:

- Enhanced surveillance of pregnant women, including the early detection of infectious diseases such as syphilis, HIV and DEN/CHIK/ZIKV.
- Implementation of communication strategies, that is, prevention and control measures against the bite of the mosquito vectors, family-planning strategies.
- Determination of the frequency of structural anomalies detected prenatally in the fetus, umbilical cord and the placenta of pregnancies affected by infection with ZIKV.
- Detection of ZIKV, and other arboviruses/viruses in products of abortion, in-utero and early newborn.
- Implementation of the strategy "Casas a prueba de *Aedes aegypti*/*Aedes aegypti*-proof houses" in Yucatan (protecting homes and families with door/window screening to prevent the access of mosquitoes), provision of mosquito repellents and larvicides, and education/promotion of best practices for the prevention of infection with DEN, CHIK and ZIKV.
- Conducting an anthropological research of sociocultural factors associated with Zika virus.

2. Enhanced surveillance of pregnant women, including the early detection of infectious diseases such as syphilis, HIV and DEN/CHIK/ZIKV

In Merida, a strong collaborative work with local authorities is already in place with interventions targeting high-risk areas and vulnerable populations. These efforts were initially organized for DEN control due to its endemic transmission in the area. Due to the introduction of CHIK in early 2014, and the most recent introduction of ZIKV, the program has extended to address these three ABD with major emphasis in ZIKV infection due to its high impact in the newborn. "Familias sin Dengue project" (**Figure 1**) (financed by SANOFI-Pasteur) has established a platform for the surveillance ad early detection of dengue infection that will be used to improve detection of ZIKV infection in target women.

2.1. Surveillance of ZIKV and urban *Aedes*-borne diseases (ABD) with in the cohort study "Familias sin dengue"

For Merida, the team has access to datasets containing the residential address of ~25,000 DEN and ~5000 CHIK cases reported during 2011–2016 including the study area. The Cohort study: "Familias sin Dengue" also provides a group of women in reproductive age who will be follow through house visits in order to detect suspicion of ZIKV infection.

ZIKV infection in the general population presents with a mild disease that remit in 3–7 days. ABD are endemic in Yucatan, and congenital infections result from the transmission of

Figure 1. Identification logo of the project families without dengue.

infection from the mother to the foetus during pregnancy. Therefore, in populations with an epidemiological association of ZIKV, pregnant women should be intentionally evaluated for ZIKV during the development of clinical manifestations similar to those described [11]. It is essential to establish a correct diagnosis in the pregnant woman due to the poor information generated on the transmission, pathophysiology and diagnosis of the impact that this infection has on the product of gestation [10, 21].

The virus in addition to being identified in saliva, blood, semen, urine, in breast milk, has also been found, but it is not known, if breastfeeding can affect the baby. But if vertical transmission has been demonstrated through amniocentesis, with confirmation the presence of ZIKV by PCR and reported lesions vary according to the time of infection and time of pregnancy. However, information about ZIKV infection proximate to the conception period is scarce. Women who are positive should have a follow-up for the search for morphological abnormalities in their products. Zika infection in the mother with transmission to the fetus, causing a group of structural abnormalities of at least the central nervous system (CNS), which produces a sequence of cerebral disruption and causes a functional disability secondary to damage, is denominated congenital Zika syndrome (CZS) [13, 22–24].

Enhanced surveillance searches for prenatal and/or neonatal morphological abnormalities in all products of pregnancies affected by ZIKV. This performed by perinatologists specialized in fetal medicine and geneticists from the dysmorphological approach. A family history is made of all the patients with the purpose of obtaining a history of genetic diseases or the use of teratogens that may imply in the diagnosis of maternal-fetal health.

According to the guidelines established by the CDC, there is no optimal time to perform an ultrasonographic screening to detect microcephaly. On a daily basis, ultrasound surveillance to detect structural abnormalities in a pregnancy is performed among the 18–20 weeks of gestation [25]. Ultrasonographic follow-up should be performed from the beginning of the ZIK symptomatology in the mother with a monthly monitoring depending on the prenatal manifestations found. The variables investigated are intrauterine growth, cephalic perimeter, central nervous system abnormalities and other structural abnormalities as well as the presence of intracerebral and/or placental calcifications and the amount of amniotic fluid.

Upon ultrasonographic findings of malformations in the product, the patient will be informed, who will decide whether or not to follow the pregnancy. In case of a request for the interruption of the pregnancy the case will be evaluated by a Bioethics Committee to define the resolution.

Pregnancies with or without ZIKV infection that culminate in abortion or early foetal death, with prior informed consent, will be performed a genetic clinical evaluation, with collection of body images, and evisceration or autopsy will be performed, as the case may be, to obtain brain tissue, hepatic, cardiac and renal, to perform histopathological study and RT-PCR for ZIKV nuclei acid in frozen and fixed tissue. The mother will be informed of the findings and genetic counseling on care and risk will be given to reduce birth defects in subsequent pregnancies. To date, there is no contraindication to vaginal delivery and breastfeeding.

2.2. Studies for the core group

The core group of the study are the pregnant women of two interventions; the cohort study (**Figure 2**): "families without dengue" (**Figure 1**) and "*Aedes aegypti*-proof houses" (**Figure 3**) in which previous informed and signed consent, with clarity the commitments that the group of professionals, as well as the participant and their family have.

When a pregnant woman, her partner and family enter the project, the following diagnostic tests and complementary studies are carried out (**Figure 4**):

A. Molecular detection of ZIKV is performed with the use of the Trioplex Real time RT-PCR Assay from the Centers for Disease Control and Prevention (CDC), for detection and differentiation of RNA from DEN, CHIK and ZIKV in serum and urine, to all members of the family.

B. Serum and urine studies with Trioplex RT-PCR monthly to the pregnant woman and her partner throughout pregnancy.

C. HIV and VDRL (syphilis) tests on the pregnant woman and her partner.

D. TORCH (toxoplasma, rubeola, cytomegalovirus and herpes) to the pregnant.

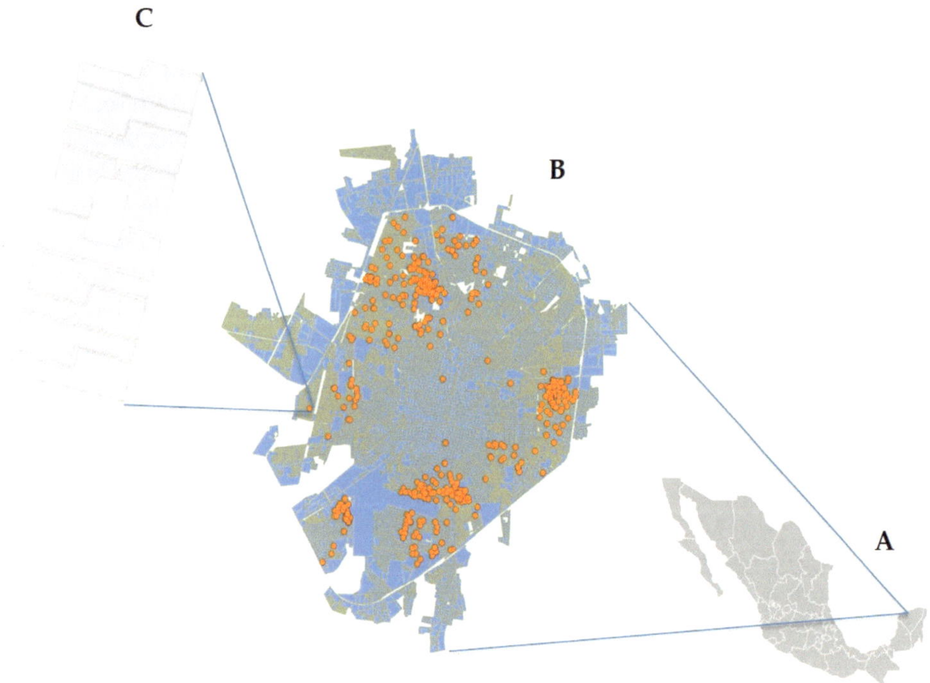

Figure 2. Map of the study areas in Merida, Yucatan, Mexico. (A) Map of Mexico with the location of the city Merida; (B) distribution of the participants of the cohort from the project families without dengue; (C) intervention area in the neighborhood Juan Pablo II.

http://dx.doi.org/10.5772/intechopen.71504

E. Blood count, blood chemistry, general urine test and erythrocyte folate profile in the pregnant woman.

F. Obstetric ultrasound (quarterly).

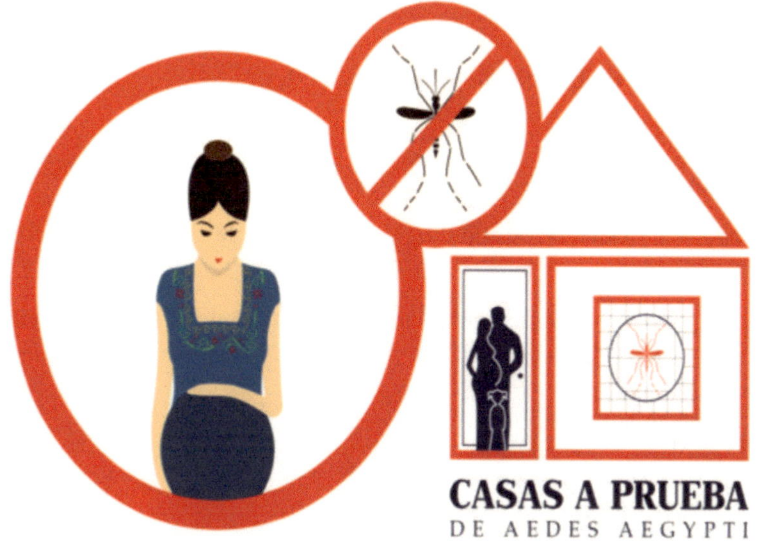

Figure 3. Identification logo of the intervention '*Aedes aegypti-proof houses*'.

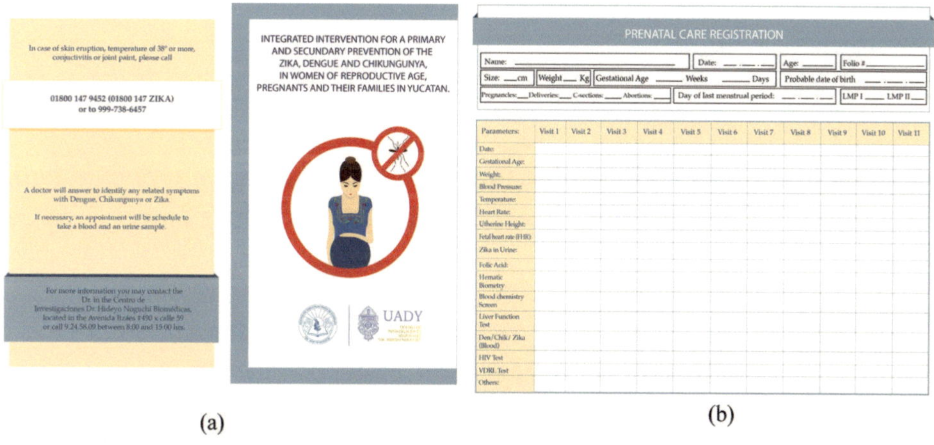

In case of skin eruption, temperature of 38° or more, conjuctivitis or joint paint, please call

01800 147 9452 (01800 147 ZIKA)
or to 999-738-6457

A doctor will answer to identify any related symptoms with Dengue, Chikungunya or Zika.

If necessary, an appointment will be schedule to take a blood and an urine sample.

For more information you may contact the Dr. in the Centro de Investigaciones Dr. Hideyo Noguchi Biomédicas, located in the Avenida Itzáes #490 x calle 59 or call 9.24.58.09 between 8:00 and 15:00 hrs.

INTEGRATED INTERVENTION FOR A PRIMARY AND SECUNDARY PREVENTION OF THE ZIKA, DENGUE AND CHIKUNGUNYA, IN WOMEN OF REPRODUCTIVE AGE, PREGNANTS AND THEIR FAMILIES IN YUCATAN.

UADY

(a)

PRENATAL CARE REGISTRATION

Name: ______ Date: __.__.__ Age: ____ Folio #____
Size: ___cm Weight___ Kg Gestational Age ____ Weeks ____ Days Probable date of birth __.__.__
Pregnancies:___Deliveries:___ C-sections:___ Abortions:___ Day of last menstrual period: __.__.__ LMP I ___ LMP II___

Parameters:	Visit 1	Visit 2	Visit 3	Visit 4	Visit 5	Visit 6	Visit 7	Visit 8	Visit 9	Visit 10	Visit 11
Date:											
Gestational Age:											
Weight:											
Blood Pressure:											
Temperature:											
Heart Rate:											
Utherine Height:											
Fetal heart rate (FHR):											
Zika in Urine:											
Folic Acid:											
Hematic Biometry											
Blood chemistry Screen											
Liver Function Test											
Den/Chik/ Zika (Blood)											
HIV Test											
VDRL Test											
Others:											

(b)

Figure 4. Carnet for pregnant women: (a) front cover with general information of the project and how to contact the team including a free telephonic number to communicate in case of presenting symptoms compatible with DEN/ CHIK/ ZIKV and (b) record for diagnostic tests and complementary studies that are recommended for pregnant women.

2.3. Positive cases for ZIKV

In case the participant or any member of the family has signs and symptoms compatible with DEN/CHIK/ZIKV, and have positive test result of: diagnostic test Trioplex RT-PCR DEN, CHIK and ZIKV and serological tests of NS1 and IgG for dengue, IgM DEN, CHIK and ZIKV [20].

At the moment that the pregnant woman presents a positive result to ZIKV by the Trioplex RT-PCR assay for DEN, CHIK and ZIKV, she is informed of the diagnosis with counselling and psychological support in case of requesting it and the following studies and assessments are requested:

A. Liver function tests.

B. Structural ultrasound/obstetric ultrasound (monthly).

C. Genetic consultation.

D. Obstetric consultation.

E. Amniocentesis with karyotype and diagnostic test Trioplex RT-PCR assay for DEN, CHIK and ZIKV, if required.

At the end of the pregnancy, vaginal, cesarean or abortion, biological samples will be collected to perform Trioplex RT-PCR assay for DEN, CHIK, ZIKV and anatomopathology and specialised studies (karyotype):

A. Maternal blood

B. Product blood

C. Amniotic fluid

D. Placenta

E. Abortion product

F. Cerebrospinal fluid

3. Implementation of communication strategies, that is, prevention and control measures against the bite of the mosquito vectors, family-planning strategies

3.1. Preventive kits

All pregnant women will be provided with an educational/preventive kit containing (**Figure 5**):

1. Proven repellent (DEET 30%) recommended by CENAPRECE (Ministry of Health) for protection outside the home.
2. Organic and environmentally friendly larvicide recommended by CENAPRECE (SPINOSAD).
3. Educational brochure for the promotion of good practices to avoid the risk of mosquito bites (including use of repellent) and for the elimination and control of Aedes breeding grounds (including the use of larvicides).
4. Thermometer for fever monitoring.

5. Personal symptom monitoring card compatible with ZIKV and Carnet for registration of laboratory tests.
6. Condoms (for prevention of possible transmission by sexual practices).
7. Access to 01800 CERO ZIKA (01800 09452) for report in case the pregnant or someone in your family presents symptoms of Zika: erythema, fever, conjunctivitis or joint pain.

To supplement these direct actions, text messages are periodically sent with information relevant to the case and reinforce preventive measures and home visits in search of febrile illnesses. Those women residing in the area where the "*Aedes aegypti*-proof houses" strategy will be evaluated will also receive the installation of long-lasting insecticide nets permanently installed in doors and windows in their homes (see more detail in the next section).

3.2. Educational intervention with women of reproductive age and their partners for the prevention of vector-borne diseases and sexually transmitted infections

The strategy of educational intervention is performed in the form of workshops. A "a pedagogical strategy, apparently simple, that by a learning methodology by doing in group, allows to build meaning to those" someone "who participate in it in order to learn and know" something ", from the integral insertion in the process" [26].

It is directed to the pregnant women who participate in the project and the women who are at reproductive age within the family nuclei. Workshops are also held with their partners so that

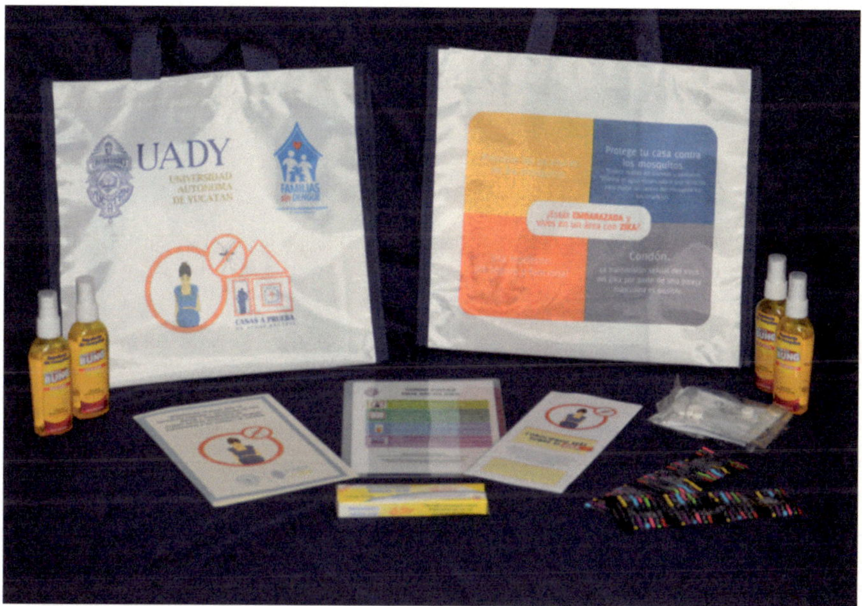

Figure 5. Prevention kits for pregnant women.

they are also aware of the information and are involved in the corresponding care during pregnancy. Basically, workshops deliver topics related to health, body, sexuality, pregnancy and strategies for prevention of vector-borne diseases and sexually transmitted infections.

The themes and objectives of the workshop are:

Topic 1. Body, pregnancy and sexuality.

General objective: to recognize the characteristics of the male and female body that make possible a pregnancy and the processes of a normal pregnancy, to identify the benefits of a planned pregnancy, the repercussions of an unplanned pregnancy, the health risks during pregnancy and to recognize the importance of prenatal care.

Topic 2. *Aedes*-borne diseases and the presence of ZIKV in pregnancy.

General objective: to provide information for the prevention of DEN, CHIK and ZIK, as well as the impact of this during pregnancy, identify the mechanisms of transmission:

A. Symptoms and signs of dengue, chikungunya and Zika.

B. Identify the activities they do and can do at home and in the community to prevent Aedes-borne diseases.

C. Impact of Zika infection on pregnancy. Congenital malformations.

D. Promotion of the use of condoms as a method of prevention of Zika.

E. Recognize the importance of preventing Zika during pregnancy.

F. Recognize Zika as a preventable disease and not as a determinant for the presence of congenital malformations.

Topic 3. HIV/STIs prevention.

Course objective: identify the risks of STI contagion with an emphasis on HIV.

A. Myths and realities.

B. Describe the most frequent STIs: HIV, gonorrhea, syphilis, herpes, and human papilloma virus.

C. Mention what are the risk practices and routes of HIV and STI transmission.

D. Describe clinical picture (differences between HIV/AIDS and STIs) and diagnostic methods, treatment and prognosis.

E. Identify methods of prevention of STIs (male condom, female condom, and abstinence.

Topic 4. Protected sex.

Course objective: Provide information on the male and female condom, and promote its use.

A. Perception of risks of contracting STIs.

B. Myths and realities.

C. Knowledge and importance of protected sex.

D. Promote the practice of sex protected in sexually active population.

E. Knowledge, characteristics and correct use of the male and female condom.

F. Explain the correct use of the female condom (steps for its use).

Topic 5. Self-care of health.

General objective: To reflect on the self-care practices of health, identifying the consequences, and benefits of self-care.

4. Determination of the frequency of structural anomalies detected prenatally in the foetus, umbilical cord, and the placenta of pregnancies affected by infection with ZIKV

Gestation products that come to term should be evaluated for a possible congenital infection by ZIKV. Serum, placenta tissue, and urine should be obtained. The search for ZIKV RNA, immunoglobulin M and neutralizing ZIKV antibodies as well as neutralizing antibodies and M immunoglobulins for DEN is indicated. Tissue collection should be done within the first 2 days of life, if possible [11]. For this reason, all pregnancy products are evaluated by a group of experts, including clinical geneticists and pathologists, and the tissues are collected and evaluated within the first 12 hours of life. The tissues are sent to the pathology service for histopathological evaluation of the placenta and the umbilical cord. Immunohistochemical staining is performed in fixed tissue and ZIKV RNA is sought by RT-PCR in fixed and frozen tissue.

Clinical evaluation implies a dysmorphlogical approach and records gestational age, Apgar, occipitofrontal perimeter, height and weight. Neurological, craniofacial features, cutaneous, thoracic, and abdominal abnormalities are recorded. If a specific abnormality is observed, an evaluation by the appropriate specialist is requested [14]. Complementary studies are carried out such as transfontanellar ultrasound, auditory, and neonatal screening.

Long-term follow-up involves evaluation for 18 months and includes follow-up by geneticist, pediatrician, pediatric ophthalmology, audiology and pediatric neurology. A medical record of occipitofrontal circumference (OFC), growth, weight, height and neuromotor development is kept [16].

The possible clinical scenarios to evaluate are:

1. The patient who is observed without microcephaly or morphological alterations and has normal neurological development being the mother of a child without or with infection by ZIK and

2. Those that are observed with microcephaly, morphological alterations and / or alterations of the neuromotor development being children of mothers without or with infection or ZIKV.

In the latter group, previously well-established genetic syndromes, added teratogenic causes and/or complementary genetic studies would be ruled out according to the individual needs of each one.

The information will be collected in a database to be able to determine the frequencies.

Patients who undergo microcephaly and/or brain disruption sequence will be given skull CT and brain MRI to define CNS damage and its relationship to congenital infection by ZIKV. Other interventions will be required according to the needs of each individual [27]. Other interventions will be required according to the needs of each individual, such as the evaluation by a paediatric cardiologist in the assumption of a cardiac murmur or paediatric nephrologist for renal morphological abnormalities that compromise the function.

5. Effective vector control interventions for integrated prevention/control of *Aedes*-borne diseases (ABD) including DEN/CHIK/ZIKV and targeted on vulnerable population (pregnant women)

As described previously, the project is working to ensure that vulnerable population (pregnant women and their families) are provided with the best possible support and care, including the access of the best supplies for vector control and the prevention of *Aedes*-borne diseases (ABD); this is personal and family protection in addition to the traditional activities performed by the local health programs. This was urged in the Zika strategic response framework and joint operations plan emitted by WHO [28], commending a proactive special care for pregnant women, such as giving adequate repellent lotion and treated mosquito nets.

5.1. How can pregnant women protect themselves from mosquito bites?

The best protection from ZIKV is preventing mosquito bites. As stated by the most important agencies worldwide such as the World Health Organization [29], the CDC [30], and the Ministry of Health in Mexico, this can be done effectively and individually (and at the family level) by the integrated use of: insect repellents, the use of physical barriers such as screens on doors and windows to prevent mosquito bites outside and inside home; in an integrated manner with the elimination of mosquito breeding sites, the application of larvicides and application of insecticides to kill adult mosquitoes to control vector populations in and around homes [31].

5.2. The entomological tools and interventions selected

For an effective and integrated prevention/control of ABD in this project we have chosen: the best repellent with diethyltoluamide (DEET) available in the local market, the intervention called "*Aedes aegypti*-proof houses" ("Casas a prueba de *Aedes aegypti*" in Spanish) which involves insecticide-treated house screening (ITS) with use long-lasting insecticidal nets (LLIN) permanently fitted to windows and doors to exclude *A. aegypti*, from houses, and the provision of an environmentally safe biorational larvicide.

5.3. Repellent

In order to select the best repellent for this project, we assessed the efficacy of commercial repellents available in Yucatan against *A. aegypti*. First results were reported by Uc-Puc et al. [32]. Protection time was determined based on WHO/CTD/WHOPES/IC and Mexican regulations (NOM-032-SSA2–2014) [32–34]. Two products, both with DEET (N,N-diethyl-3-methylbenzamide) >25% met the recommended protection (≥6 hours). The best repellent was "Stop fly bung" with DEET at 30%. Efficacy was directly proportional to the concentration of DEET; botanicals repellents resulted no protective. Repellents with DEET provided more protection against *A. aegypti* and botanical repellents, including impregnated wristbands provided no protection [33, 34].

5.4. Larvicide

The selection of the larvicide was supported by previous evaluations of its efficacy under laboratory and field conditions and in consistency with the list of available products recommended by the Mexican Ministry of Health [35–38]. The product selected was the biorational and environmentally friendly larvicide Natular® DT (Spinosad 7.48%; Clarke Mosquito Control, IL, USA; WHOPES approved) which is available in the formulation of a tablet for containerized water. This product and formulation is a highly effective larvicide against *A. aegypti* with a residuality of 9 weeks in field studies. In addition is non-toxic to humans, and also an option for the management of insecticide resistance for local *A. aegypti* populations.

5.5. *Aedes aegypti*-proof houses

The intervention called *Aedes aegypti*-proof houses ("Casas a prueba de *Aedes aegypti*") involves insecticide-treated house screening (ITS) with use long-lasting insecticidal nets (LLIN) permanently fitted to windows and doors to exclude *A. aegypti*, vector of dengue, chikungunya and Zika (**Figure 6**).

The selection of these interventions was based on previous studies of our group. Briefly, from 2009 to 2014, cluster randomized controlled trials were conducted in the Mexican cities of Acapulco and Merida. Intervention clusters received *Aedes aegypti*-proof houses with ITS and followed-up during 2 years. Overall, results showed significant and sustained reductions on adult vector densities in houses in the treated clusters with ITS after 2 years: ca. 50% on the presence ($OR \leq 0.62$, $P < 0.05$) and abundance ($IRR \leq 0.58$, $P < 0.05$) of indoor-resting adults. ITS

Figure 6. *Aedes aegypti*-proof houses.

on doors and windows are 'user-friendly' tool, with high levels of acceptance, requiring little additional work or behavioral change by householders. ITS is a housing improvement that should be part of the current paradigms for urban vector-borne disease control [36].

As described in Che-Mendoza et al. [37] and Manrique-Saide et al. [39], Duranet® screens (0.55% w.w. alpha-cypermethrin-treated non-flammable polyethylene netting [145 denier; mesh1⁄4132 holes/sq. inch]; Clarke Mosquito Control, Roselle, IL, USA; WHOPES approved for LLIN use) were mounted in aluminum frames custom-fitted to doors and windows of houses in collaboration with a local small business [37, 40]. An average of two doors and five windows by house were installed in each intervention cluster. During installation, at least one person in every household received information from research staff about the proper use and maintenance of ITS.

5.6. Government agency as key partner in the intervention

These tools and interventions have fully political support by the Government of the State of Yucatán who also gave financial support. In Merida, a strong collaborative work with local authorities is already on-going and with this information we are already starting interventions (and will start others) targeting high-risk areas and vulnerable populations. The authorities of Mexico are considering how to expand *Aedes*-proof housing to as many homes as possible, probably as a targeted intervention for high-risk areas (hot-spots) of endemic localities.

6. Zika, culture and pregnancy: An anthropological overview

The particularity of Zika virus presents a 'double identity' because it belongs to mosquito vector-borne diseases and sexually transmitted infections. The initial cultural meanings of ZIKV produced in local population uncertainty, misinformation, social fears and rumours. Even in social networks like Facebook or Twitter, there was an entanglement of discourses that increased confusion about symptoms, treatments, and consequences of this new outbreak. Currently, we have noticed that incommensurability continues at stake on the cultural dimension of this emergent disease. That's the reason why an anthropological overview will help us to a better understanding of the sociocultural determinants of ZIKV. The final purpose is to address cultural barriers and social acceptance of an integrated prevention model that pursuit improves population health in Yucatan.

6.1. Research process

Our group of anthropologists is currently conducting three studies as part of this integrated prevention model. First one, there is a research focused on sociocultural factors that pregnant women associated with Zika virus in Yucatan, Mexico. Ethnographic and in-deep interview methods are used for the collection data. The guidelines of this study are: source of information about ZIKV; the mechanisms associated with Zika infection; social perceptions of symptoms; cultural beliefs about microcephaly and Guillain-Barre Syndrome; risk perceptions on Zika virus; and the prevention and control practices of this target population. Also, we make anthropological analysis on focus group and workshops activities developed by medical crew with pregnant women.

Second, there is research which main goal is focused on human mobility and its relation with DENV, CHIKV and ZIKV transmission in Yucatan. A former and pioneer study was deployed in Iquitos, Peru since 2007 [40]. We adapted surveys applied there to Yucatecan context, coordinated by Gonzalo Vazquez-Prokopec. There are two types of questionnaires: the retrospective movement survey and the prospective movement survey. The first one, will help to identify the amount of time DEN/CHIK/ZIKV-infected person spends at home as well array locations. The second one is for not infected individuals. In both cases, we want to know about their routine movements and the time-spatial quantification of them.

Third, we are making a rapid assessment on social perception about insecticide-treated screening (*Aedes aegypti*-proof houses). Pregnant women and their families are the main target of this study. The topics this survey are: acceptance of the intervention, installation process, family perception on positive and negative impacts of this technological innovation, comfort and heat perception, families' opinion on scale-up this project.

Finally, we are conducting a sociological assessment on the integrative prevention model. We seek to address what is the social perception of pregnant women about the intervention (enrollment process, information of Zika and ABD prevention practices), and follow-up health care of pregnancy). In addition, we are exploring with in-deep interviews vulnerability experiences that pregnant women have lived during and after the emergence of Zika outbreak in Yucatan.

6.2. Anthropological contributions

We expect, that this three and incoming social studies, will provide cross-cultural perspectives on entomological, medical, ecological and epidemiological approaches. The common goal to achieve, as a multidisciplinary team, is to produce a very positive impact of named interventions in collaboration with local communities and government institutions. Since anthropological point of view, we are strengthening the integrated prevention model by studying how human factor is interweaved in an eco-bio-social context.

7. Discussion and conclusions

Here we described and discussed the design and current implementation of an integrated model for primary and secondary preventive interventions for ZIKV, along with other ABD and sexually transmitted infections that impact reproductive health, pregnancy and perinatal life stages in women and their upspring in Merida, Mexico.

This intervention is being developed in a population of 10,000 people of the city of Merida and with the participation of a multidisciplinary group of public health professionals (epidemiologists, obstetricians, geneticists, virologists, immunologists, pathologists, entomologists, ecologists, sexologists and anthropologists), in collaboration with the Ministry of Health and the Government of Yucatan.

There is not a full understanding of the risk for pregnant women, and the teratogenicity of prenatal infections with ZIKV has not been completely elucidated [7, 10, 11, 13–15]. It is urgent that a protocol designed for surveillance of pregnant women in endemic and invasive zones of *A. aegypti* allow collection of information about the effects of ZIK in pregnant women, and to delineate the neonatal phenotype of congenital abnormalities of the fetus related with ZIK. This protocol will allow the establishment strategies related to the prevention and control of emerging infectious diseases for future use [41, 42].

The implementation of complementary, innovative (not included by the traditional vector control programs), feasible and locally adapted approaches to vector control to reduce the risk of Zika and other ABD, through personal protection, environmental management and community-based partnership models can improve the current strategies for ABD prevention and control.

The results from the surveillance in a vulnerable population, will enable at least four strategies: (1) create a program focused on a rapid recognition of any clinical symptoms for ZIKV; (2) set up rehabilitation programs for motor, hearing and visual disabilities in order to support neurodevelopment and quality of life for patients; (3) determine the performance of current vector control and (4) establish measures and protocols in accordance with each specific population to ensure the best results in surveillance, prevention and control of vector-borne diseases in vulnerable populations. Finally, this strategy will allow development of human resources in research, including educational cooperation and special experience in handling patients affected by ZIKV.

Therefore, this proposal will have direct impact on social, educational, economic and environmental aspects concerned with the general health of the population, the collaboration and engagement with government institutions are key factors to accomplish common goals on public health and preventive strategies against vector-borne diseases.

Acknowledgements

The studies are funded by the Office of Infectious Disease, Bureau for Global Health, U.S. Agency for International Development, under the terms of an Interagency Agreement with CDC. The opinions expressed herein are those of the author(s) and do not necessarily reflect the views of the U.S. Agency for International Development. It is also financed by Sanofi Pasteur Sanofi Pasteur INC, Special Programme for Research and Training in Tropical Diseases (TDR) at the World Health Organization in a partnership with the Ecosystem and Human Health Program of the International Development Research Centre (IDRC) of Canada within the program "Towards Improved Dengue and Chagas Disease Control through Innovative Ecosystem Management and Community-Directed Interventions: An Eco-Bio-Social Research Programme on Dengue and Chagas Disease in Latin America and the Caribbean" (Project 104951-001) and the Canadian Institutes of Health Research (CIHR) and IDRC (Preventing Zika disease with novel vector control approaches Project 108412). The nets, repellent and larvicide employed in this study were donated by the company Public Health Supply and Equipment de Mexico, S.A. de C.V. The nets and larvicide and repellent employed in this study were donated by the company Public Health Supply and Equipment de Mexico, S.A. de C.V.

Author details

Norma Pavía-Ruz[1], Silvina Contreras-Capetillo[1], Nina Valadéz-González[1], Josué Villegas-Chim[1], Rafael Carcaño-Castillo[1], Guillermo Valencia-Pacheco[1], Ligia Vera-Gamboa[1], Fabián Correa-Morales[2], Josué Herrera-Bojórquez[1], Hugo Delfín-González[1], Abdiel Martín-Park[3], Guadalupe García Escalante[1], Gonzalo Vázquez-Prokopec[4], Héctor Gómez-Dantés[5] and Pablo Manrique-Saide[1]*

*Address all correspondence to: msaide@correo.uady.mx

1 Universidad Autónoma de Yucatán, Mérida, Yucatán, México

2 Centro Nacional de Programas Preventivos y Control de Enfermedades (CENAPRECE), Secretaria de Salud, Mexico

3 Cátedras-CONACYT Universidad Autonoma de Yucatan, Merida, Yucatan, Mexico

4 Emory University, Atlanta, USA

5 Instituto Nacional de Salud Publica, Cuernavaca, Morelos, Mexico

References

[1] Hatnagar J, Rabeneck DB, Martines RB, Reagan-Steiner S, Ermias Y, Estetter LBC, Suzuki T, Ritter J, Keating MK, Hale G, Gary J, Muehlenbachs A, Oduyebo T, Meaney-Delman D, Shieh W, Zaki SR. Zika virus RNA replication and persistence in brain and placental tissue. Emerging Infectious Diseases. 2017;**23**(3). DOI: 10.3201/eid2303.161499

[2] Duffy MR, Chen TH, Hancock WT, Powers AM, Kool JL, Lanciotti RS. Zika virus outbreak on Yap Island, Federated States of Micronesia. The New England Journal of Medicine. 2009;**360**:2536-2543

[3] Musso D, Nhan T, Robin E, Roche C, Bierlaire D, Zisou K, Shan Yan A, Cao-Lormeau VM, Broult J. Potential for Zika virus transmission through blood transfusion demonstrated during an outbreak in French Polynesia, November 2013 to February 2014. Euro Surveillance. 2014;**19**(14):1-3

[4] Faye O, Freire CCM, Iamarino A, Faye O, Oliveira JVC, Diallo M, et al. Molecular evolution of Zika virus during its emergence in the 20th century. PLoS Neglected Tropical Diseases. 2014;**8**(1):e2636

[5] Haddow AD, Schuh AJ, Yasuda CY, Kasper MR, Heang V, Huy R, Guzman H, Tesh RB, Waver SC. Genetic characterization of Zika virus strains: Geographic expansion of the Asian lineage. PLoS Neglected Tropical Diseases. 2012 Feb;**6**(2):1-7

[6] Pan American Health Organization/World Health Organization. Zika Suspected and Confirmed Cases Reported by Countries and Territories in the Americas Cumulative

Cases, 2015–2017Updated as of 22-06-2017. Washington, D.C.: PAHO/WHO; 2017 Pan American Health Organization.www.paho.org. (c) PAHO/WHO, 2017

[7] Pacheco O, Beltrán M, Nelson C, Valencia D, Tolosa N, Farr SL, et al. Zika virus disease in Colombia—Preliminary report. The New England Journal of Medicine. 2016 Jun 15. p. 1-16. DOI: 10.1056/NEJMoa1604037

[8] Boletín Epidemiológico Sistema Nacional de Vigilancia Epidemiológica Sistema Único de Información. http://www.gob.mx/salud/documentos/casos-confirmados-de-infeccion-por-virus-zika-2017.pdf [Acceso, Junio 26, 2017]

[9] Sistema de Vigilancia Epidemiológica de Enfermedad por ZIKA. http://www.gob.mx/salud/documentos/casos-confirmados-de-infeccion-por-virus-zika-2017. [Acceso, Junio 26, 2016]

[10] Brasil P, Pereira JP, Moreira ME, Ribeiro Nogueira RM, Damasceno NG, Wakimoto M, Rabello RS, Valderramos SG, Halai UA, et al. Zika virus infection in pregnant women in Rio de Janeiro. The New England Journal of Medicine. 2016;**375**:2321-2334

[11] Fleming-Dutra KE, Nelson JF, Fischer M, Saples JE, Karwowski MP, Mead P, Villanueva J, Renquist CM, Minta AA, Jamieson DJ, Honein MA, Moore CA, Rasmussen SA. Update guidance for health care providers caring for infants and children with possible zika virus infection-United States, February 2016. Morbidity and Mortality Weekly. 2016;**65**(7):182-187. DOI: http://dx.doi.org/10.15585/mmwr.mm6507e1

[12] Oliveira Melo AS, Malinger G, Ximenes R, Szejnfeld PO, Sampaio A, Bispo de Filippis AM. Physician alert. Zika virus intrauterine infection causes feltal brain abnormality and microcephaly: Tip of the iceberg? Ultrasound in Obstetrics & Gynecology. 2016;**47**:6-7

[13] Mlakar J, Korva M, Tul N, Popovic M, PoljsakçPrijatelj M, Mraz J, Kolenc M, Rus KR, Vipotnik TV, Vodušek VF, Vizjak A, Pižem J, Petrovec M, Županc TA. Zika virus associated with microcephaly. The New England Journal of Medicine. 2016;**374**:951-958. DOI: 10.1056/NEJMoa1600651

[14] Moore CA, Staples JE, Dobyns WB, Pessoa A, Ventura CV, Borges da Fonseca E, Marques Ribeiro E, Ventura LO, Neto NN, Arena JF, Sonja A, Rasmussen A. Characterizing the pattern of anomalies in congenital Zika syndrome for pediatric clinicians. JAMA Pediatrics. 2016:E1-E8. DOI: 10.1001/jamapediatrics.2016.3982

[15] Campo M, Feitosa IM, Ribeiro EM, et al. The phenotypic spectrum of congenital Zika syndrome. American Journal of Medical Genetics. 2017;**173**:841-857

[16] Reynolds MR, Jones AM, Petersen Em Lee EH, Rice ME, Bingham A, Ellingtos SR, Evert N, et al. Vital signs: Update on Zika virus-associated birth defects and evaluation of all U.S. infants with congenital Zika virus exposure—U.S. Zika pregnancy. MMWR. 2017;**66**:1-9

[17] Reusken C, Pas S, Geurts van Kessel C, Mögling R, van Kampen J, Langerak T, Koopmans M, van der Eijk A, van Gorp E. Longitudinal follow-up of Zika virus RNA in semen of a traveller returning from Barbados to the Netherlands with Zika virus disease. Eurosurveillance. 2016;**21**(23). DOI: 10.2807/1560-7917.ES.2016.21.23.30251

[18] Nicastri E, Castilletti C, Liuzzi G, Lannetta M, Capobianchi MR, Lppolito G. Persistent detection of Zika virus RNA in semen for six months after symptom onset in a traveller returning from Haiti to Italy. Eurosurveillance. 2016;**21**(32) pii=30314. DOI: http://dx.doi.org/10.2807/1560-7917.ES.2016.21.32.30314

[19] Landry ML, St George K. Laboratory diagnosis of Zika virus infection. Archives of Pathology & Laboratory Medicine. 2017;**141**(1):60–67. DOI: http://dx.doi.org/10.5858/arpa.2016-0406-SA

[20] Rabe IB, Staples EJ, Villanueva J, Hummel KB, Johnson JA, Rose L, Hills S, Wasley A, Fischer M, Powers AM. Interim guidance for interpretation of Zika virus antibody test results. Morbidity and Mortality Weekly Report. 2016;**65**(21):543-546. DOI: http://dx.doi.org/10.15585/mmwr.mm6521e1

[21] Besnard M, Lastere S, Teissier A, Cao-Lormeau V, Musso D. Evidence of perinatal transmission of Zika virus, French Polynesia, December 2013 and February 2014. Eurosurveillance. 2014;**19**(13):1-4. DOI: http://dx.doi.org/10.2807/1560-7917.ES2014.19.13.20751

[22] Calvet G, Aguiar RS, Melo ASO, Sampaio SA, de Filippis I, Fabri A, Araujo ESM, de Sequeira PC, Mendonça MCL, de Oliveira L, Tschoeke DA, Schrago CG, Thompson FL, Brasil P, dos Santos FB, Nogueira RMR, Taruni A, Filippis AMB. Detection and sequencing of Zika virus from amniotic fluid of fetuses with microcephaly in Brazil: A case study. Lancet Infectious Diseases. 2016;**16**(6):653-660. DOI: http://dx.doi.org/10.1016/S1473-3099(16)00095-5

[23] Musso D, Roche C, Romin E, Nhan T, Teissier A, Cao-Lormeau V. Potential sexual transmission of zika virus. Emerging Infectious Diseases. 2015;**21**:359-361. DOI: 10.3201/eid2102.141363

[24] Sarno M, Sacramento GA, Khouri R, do Rosairo MS, Costa F, Archanjo G, Santos LA, Nivison N Jr, Vasilakis N, Ko AI, de Almeida ARP. Zika virus infection and stillbirths: A case of Hydrops Fetalis, Hydranencephaly and fetal demise. PLoS Neglected Tropical Diseases. 2016;**10**(2):e0004517. DOI: 10.1371/journal.pntd.0004517

[25] Carvalho FH, Cordeiro KM, Peixoto AB, Tonni G, Moron AF, Feitosa FE, Feitosa HN, Junior EA. Associated ultrasonographic findings in fetuses with microcephaly because of suspected Zika virus (ZIKV) infection during pregnancy. Prenatal Diagnosis. 2016;**36**(9):882-887. DOI: 10.1002/pd.4882

[26] Andrade-Calderón MC, Muñoz-Dagua. El Taller Crítico: Una propuesta de trabajo interactivo. Tabula Rasa. 2004;**2**:251-262

[27] Greenwood M. Zika and Glaucoma Linked for First Time in New Study [Internet]. 2016. Available from: http://news.yale.edu/2016/11/30/zika-and-glaucoma-linked-first-time-new-study [Accessed: 2016-12-17]

[28] World Health Organization. Zika Strategic Response Framework & Joint Operations Plan [Internet]. 2016. Available from: http://www.who.int/emergencies/zika-virus/strategic-response-framework.pdf [Accessed: 2016-12-18]

[29] World Health Organization. Zika Virus and Complications: Questions and Answers [Internet]. Available from: http://www.who.int/features/qa/zika/en/ [Accessed: 2016-12-18]

[30] Centers for Disease Control and Prevention. Zika Virus [Internet]. Available from: https://www.cdc.gov/zika/prevention/index.html [Accessed: 2016-12-18]

[31] Petersen LR, Jamieson DJ, Powers AM, Honein MA. Zika virus. The New England Journal of Medicine. 2016;**374**:1552-1563. DOI: 10.1056/NEJMra1602113 (20)

[32] Uc-Puc V, Herrera-Bojórquez J, Carmona-Carballo C, Che-Mendoza A, Medina Barreiro A, Chablé-Santos J, Arredondo-Jiménez JI, Flores Suárez A, Manrique-Saide P. Efectividad de repelentes comerciales disponibles contra el mosquito *Aedes aegypti* (L.) vector del virus de dengue y Chikungunya en Yucatán, México. Salud Pública de México. 2016;**58**:472-475. DOI: http://dx.doi.org/10.21149/spm.v58i4.8030

[33] World Health Organization. Guidelines for efficacy testing of mosquito repellents for human skin. In: Pesticide Evaluation Scheme [Internet]. 2009. Available from: http://apps.who.int/iris/bitstream/10665/70072/1/WHO_HTM_NTD_WHOPES_2009.4_eng.pdf [Accessed: 2016-12-19]

[34] Norma Oficial Mexicana. Para la Vigilancia epidemiológica, prevención Y Control de Enfermedades Transmitidas Por Vector, NOM-032-SSA2–2014 [Internet]. 2015. Available from:http://www.cenaprece.salud.gob.mx/programas/interior/vectores/descargas/pdf/NOM_032_SSA2_2014.pdf [Accessed: 2016-12-19]

[35] Geded E, Che-Mendoza A, Medina-Barreiro A, Arisqueta-Chablé C, Dzul-Manzanilla F, Manrique-Saide P. Control of Aedes Aegypti in storm sewers with NatularTM DT and EC in Merida, Yucatan. In: Clark GG, Fernández-Salas I, editors. Mosquito vector biology and control in Latin America—A 23rd symposium. Journal of the American Mosquito Control Association. 2013;**29**(3):251-269

[36] Vazquez-Prokopec G, Lenhart A, Manrique-Saide P. Housing improvement: A renewed paradigm for urban vector-borne disease control? Transactions of the Royal Society of Tropical Medicine and Hygiene. 2016;**00**:1-3. DOI: 10.1093/trstmh/trw070

[37] Che-Mendoza A, Guillermo-May G, Herrera-Bojórquez J, Barrera-Pérez M, Dzul-Manzanilla F, Gutierrez-Castro C, Arredondo-Jiménez JI, Sánchez-Tejeda G, Vazquez-Prokopec G, Ranson H, Lenhart A, Sommerfeld J, McCall PJ, Kroeger A, Manrique-Saide P. Long-lasting insecticide treated house screens and targeted treatment of productive breeding-sites for dengue vector control in Acapulco, Mexico. Transactions of the Royal Society of Tropical Medicine and Hygiene. 2015;**109**(2):106-115. DOI: 10.1093/trstmh/tru189

[38] Centro Nacional de Programas Preventivos y Control de Enfermedades. Lista Actualizada de Productos Recomendados para El Combate de Insectos Vectores De Enfermedades a Partir de. 2016 [Internet]. Available from: http://www.cenaprece.salud.gob.mx/programas/interior/vectores/descargas/pdf/ListaActualizadaProductoRecomendadosCENAPRECE2016_1.pdf [Accessed: 2017-01-04]

[39] Manrique-Saide P, Che-Mendoza A, Barrera-Perez M, Guillermo-May G, Herrera-Bojorquez J, Dzul-Manzanilla F, Gutierrez-Castro C, Lenhart A, Vazquez-Prokopec G, Sommerfeld J, McCall P, Kroeger A, Arredondo-Jimenez JI. Use of insecticide-treated house screens to reduce infestations of dengue virus vectors, Mexico. Emerging Infectious Diseases. 2015;**21**(2):308-311. DOI: 10.3201/eid2102.140533

[40] Paz-Soldan VA, Reiner RC Jr, Morrison AC, Stoddard ST, Kitron U, Scott TW, Elder JP, Halsey ES, Koshel TJ, Astete H, Vazquez-Prokopec G. Strengths and weaknesses of global positioning system (GPS) data-loggers and semi-structured interviews for capturing fine-scale human mobility: Findings from Iquitos, Peru. PLoS Neglected Tropical Diseases. 2014;**8**(6):e2888. DOI: 10.1371/journal.pntd.0002888

[41] Faherty LJ, Rasmussen SA, Lurie N. A call for science preparedness for pregnant women during public health emergencies. American Journal of Obstetrics and Gynecology. 2017;**216**(1):34-36

[42] Lurie N, Manolio T, Patterson A, Collins F, Frieden T. Research as a part of public health emergency response. The New England Journal of Medicine. 2013;**368**:1251-1255

Chapter 5

Cocirculation and Coinfection Associated to Zika Virus in the Americas

Jorge A. Sánchez-Duque,
Alfonso J. Rodríguez-Morales, Adriana M. Trujillo,
Jaime A. Cardona-Ospina and
Wilmer E. Villamil-Gómez

Additional information is available at the end of the chapter

http://dx.doi.org/10.5772/intechopen.77180

Abstract

Zika virus, a flavivirus, has arrived to Latin America in 2013. It became evident causing epidemics since 2015, first in Brazil and later in other countries in the region, such as Colombia, with a higher peak in 2016. The World Health Organization (WHO), based on cumulated evidence on its association with Guillain-Barre syndrome (GBS) and microcephaly and other birth defects (also the congenital Zika syndrome, CZS), declared for a period of almost a year, an international public health emergency. Epidemics in the region caused around 1 million cases with also additional complications beyond GBS and the CZS, which in patients with comorbidities lead to deaths. Among the events studied in the region, a number of cases with arboviral coinfections/codetection (dengue and chikungunya) were described and published beginning in Colombia and later in Brazil. In addition to that, cocirculation and still ongoing research on antibody-dependent enhancement (ADE) are challenges for physicians and public health authorities, given the implications for clinical manifestations and serological diagnosis in patients with previous exposition to other flaviviruses. We reviewed such aspects in this chapter.

Keywords: Zika, flavivirus, dengue, chikungunya, coinfection, cocirculation, arboviruses, epidemics, epidemiology

1. Introduction

Latin America and the Caribbean (LAC) have been threatened by an unprecedented explosion of emergent and reemerging arboviral epidemic outbreaks, such as the recent Zika virus

(ZIKV) and chikungunya virus (CHIKV) epidemics, but certainly in addition to previous endemoepidemic seasons of urban dengue (DENV) and sylvatic yellow fever (YFV). These emerging viral infections are largely due to a number of factors such as climate change, levels of urbanization, migration and tourism, commercial exchange activities, susceptible geographical areas (tropical and subtropical regions), as other, that have provided an ideal blend for vector availability and virus spreading among susceptible hosts. These have allowed a spillover of these pathogens from their naturally occurring niches and reservoirs to susceptible urban settings and newly unexposed geographic areas, showing how viruses are dynamic players in the ecology of the planet, particularly in tropical and subtropical areas [1, 2].

Figure 1. Conditions prone for arboviral *Aedes* vectors proliferation. Area endemic for Zika and other arboviruses cocirculation and coinfections, La Virginia Risaralda, Colombia.

Arboviruses of medical importance (https://www.ncbi.nlm.nih.gov/Taxonomy/) are mostly included in the families Flaviviridae, Togaviridae, and Bunyaviridae. The Flaviviridae family includes dengue virus (DENV), yellow fever virus (YFV), Zika virus (ZIKV), and the Japanese encephalitis virus group (including the Japanese encephalitis virus, JEV, Usutu virus, USUV, and the West Nile virus WNV) [1–4]. The Togaviridae family includes chikungunya virus (CHIKV), Eastern equine encephalitis virus (EEEV), Mayaro virus (MAYV), Ross River virus (RRV), Sindbis virus (SINV), Venezuelan equine encephalitis virus (VEEV), and Western equine encephalitis virus (WEEV). Oropouche virus (OROV) belongs to the family Bunyaviridae [2, 3].

ZIKV shares multiple biological clinical similarities with other of those flaviviral and non-flaviviral arboviruses [1, 2]. The mosquito vectors *Aedes aegypti* and *Ae. albopictus* have facilitated these viruses' propagation; and have challenged the existing response capacities of local health systems. Many of the affected countries in Latin American countries were already facing existing challenges to build robust and reliable health systems [1, 4]. The presence of *Ae. aegypti* and *Ae. albopictus* is facilitated by socioeconomical and urban ecoepidemiological conditions, particularly the presence of old tires, where rain water is cumulated in poverty endemic areas (**Figure 1**).

In 2015, ZIKV emerged in LAC (although circulated apparently since 2013 in a silent pattern) as a leading public health priority, approximately 2 years after the CHIKV epidemic, leading to ZIKV, CHIKV, and DENV cocirculation, and the reporting of coinfections in different combinations. Although in the case of ZIKV infection, 80% of cases could be asymptomatic, clinical diagnosis of ZIKV, CHIKV, and DENV remains a challenge due to the considerable overlap in clinical presentations between them, and with other viral and nonviral infections [1, 4, 5], in addition also the problem of infection in patients with existing comorbidities that can lead to atypical, severe, and even fatal cases.

2. Cocirculation and coinfection of arboviruses

Currently, one of the major public health and biomedical challenges for the region is represented by cocirculation and coinfection of different flaviviral and nonflaviviral arboviruses [1, 5–9]. Their differentiation as their cross-reactivity between flaviviruses, particularly, implies immunological and diagnostic problems, when used serological tests [10–13].

The cocirculation of ZIKV, CHIKV, and DENV, in addition to other emerging and remerging pathogens (such as YFV, MAYV, OROV, among others), presents a number of challenges for clinical care and laboratory diagnosis in endemic areas [1, 13–21].

Patients infected with one or more of these viruses can present with similar clinical manifestations over a wide range of viremia, in some cases, suspecting that with coinfections, clinical symptoms of one of the infections can predominate while the other being asymptomatic [1, 22–24].

Clinical manifestations can initially guide the differential diagnosis and possibility of coinfection for the physicians, even in primary care, however, being challenging in multiple cases given the overlap of multiple symptoms (**Table 1**) [1].

This should be complemented with the epidemiological information of each case and epidemiology of cocirculating arboviruses, which is changing but available through multiple systems (e.g., WHO, Pan-American Health Organization [PAHO], ProMEDmail, RECOLZIKA). The broad range of possible viral and nonviral coinfecting agents and the nonspecific signs and symptoms at the initial stages of infection complicate even more the diagnostic approach to these cases, beyond the clinical aspects, implying as well the needs for the so-called multiplex diagnostic tools (specially molecular, those detecting RNA) to confirm a diagnosis of single infection or coinfection [1, 6, 7]. Recent field and clinical data have indicated that in nature, the people and mosquitoes may be infected by multiple arboviruses more frequently than has been previously appreciated, but still epidemiological studies need to approach in which proportion these occur during epidemics and nonepidemic seasons. In addition, exposure to multiple viruses tended not to affect mosquito susceptibility: mosquito midgut infection rates for each virus are similar whether the virus was delivered alone or in combination with another arbovirus [8, 11, 17]. Once a mosquito is infected, the virus disseminates through the mosquito body and replicates in the salivary gland before reaching the saliva to be transmitted [8]. In addition to this, there is no clear information yet on how is the viral interaction among infected cells, if they infected different cells, and how the immune cells in the human host as well in the mosquito interact when coinfections occur [8–15].

Clinical findings	**Main arboviruses**			
	CHIKV	**DENV**	**MAYV**	**ZIKV**
Fever	+++	++++	++++	++/0[a]
Myalgia/arthralgia	++++	+++	+++	++
Edema in limbs	0	0	0	++
Maculopapular rash	++	++	++	+++[b]
Retro-ocular pain	+	++	++	++
Conjunctivitis, nonpurulent	+	0	0	+++
Lymphadenopathies	++	++	+	+
Hepatomegaly	++	0	+	0
Leukopenia/thrombocytopenia	++	+++	++	0/+[c]
Hemorrhages	+	+++	0	0/+[c]

[a]Depends on geography and phylogeny of the virus, in some areas, patients do not have fever.
[b]Pruriginous (mild to severe).
[c]In some cases, these findings have been reported.

Table 1. Main clinical findings in CHIKV, DENV, MAYV, and ZIKV [1].

The ability of *Ae. aegypti* mosquitoes to be coinfected and cotransmit arboviruses could have important implications for the epidemiology, epidemics and syndemics (multiple simultaneous epidemics). The likelihood of coinfection by multiple *Ae. aegypti*-borne viruses may be increasing, especially during syndemics and high attack rates with multiple infected hosts. The first report of CHIKV and DENV coinfection occurred in 1964; those arboviruses were isolated from a single blood specimen taken from a patient in the acute phase of a dengue-like illness seen at Christian Medical College Hospital, Vellore, South India, in October that year [8]. The first cases of ZIKV coinfection and other arboviruses have already been reported: with DENV in New Caledonia in 2015 [9], which occurred in a traveler coinfected with ZIKV and DENV-3 and a local patient who was coinfected with ZIKV and DENV-1. In Colombia, during the beginning of ZIKV epidemics, a triple coinfection (the first globally reported), with DENV and CHIKV was reported in 2016 [10]. Also, in Colombia, in a pregnant woman, ZIKV with DENV and CHIKV was reported in the same year [11], in this case with a positive evolution.

When symptomatic infection with both ZIKV and DENV occurs, the clinical manifestations are apparently more severe [12]. However, it is not clear whether coinfection with multiple viruses could result in interspecific competition and interference during various stages of infection, or whether infection with one virus could enhance transmission of another, since our current understanding of coinfection and cotransmission by *Aedes* mosquitoes is limited [8].

Symptoms presented in reported cases of coinfections were severe arthralgia and joint swelling, nevertheless, the impact of coinfection of ZIKV and DENV on severity of illness in patients remains to be determined [12]. Recently, a study in Tolima, Colombia, following patients with post-CHIKV chronic inflammatory rheumatism (pCHIK-CIR), identified that in those patients with sequential (posterior) ZIKV infection, frequency of arthralgia and rheumatological persistent symptoms was higher [13]. The prevalence of arthralgia was higher in the group with Zika (50%) than the group without Zika post-CHIKV infection (33%). With further studies, it may be possible to prove if initial CHIKV infection is a risk factor for Zika virus disease, or if infection by other viruses after CHIKV increases likelihood of prolonging and/or intensifying the symptoms of the chronic phase. It cannot be ruled out that there are other cases of double or even triple coinfection (dengue, chikungunya, and Zika), as has been reported in Colombia in 2016 and it is also a key point to continue researching [13]. Even more, in such areas, during outbreaks, such as those occurred during 2015 for CHIKV and later for ZIKV in 2016, patients in countries such as Colombia were tested primarily for DENV, and just when negative tested for CHIKV, and if negative for DENV and CHIKV, then tested for ZIKV. This was the framework of the arboviruses sentinel network, which initially not allowed the detection of coinfections. Nevertheless, in the context of certain cases and research, clinicians were aware of infections with multiple pathogens in the differential diagnosis of dengue-like illness, especially in patients who returned from tropical and endemic areas [9–11, 14]. Coinfection diagnostic procedure could be improved by using multiplex RT-PCR, given the frequent cocirculation of multiple arboviruses in tropical regions [9–11, 14].

In Recife, northeast Brazil, the ZIKV and CHIKV coinfection has been associated with a severe case of meningoencephalitis associated with peripheral polyneuropathy in a 74-year-old patient [15], and a few cases of exanthematous illness associated with ZIKV, CHIKV, and DENV [16] also in Salvador, Bahía, Brazil have been reported. Although neurologic disturbances associated with several arboviruses were recognized, apparently there is a much higher frequency in neurological impairment following CHIKV and ZIKV infection than those resulting from DENV [15]. Nevertheless, as DENV is also neuropathogenic [17], multiple neurological consequences can occur during coinfections at this level, including central but also peripheral manifestations, including the Guillain-Barre syndrome, not only reported for ZIKV [18] but also for DENV [19] and CHIKV [20–22].

Some guidelines in USA and Brazil have been formulated to assist clinicians in assessing ZIKV in patients with DENV and CHIKV negative samples and in those who are negative on sequential testing for both pathogens. However, even if these guidelines are followed, coinfections may be missed [5, 11]. This complicates the diagnosis of patients with an acute febrile illness, as the spectra of clinical manifestations that result from infection with these viruses overlap significantly. Diagnosis is further complicated by cross-reactions observed in ZIKV-positive patients tested using immunoglobulin M (IgM) or nonstructural protein 1 (NS1) assays for DENV and vice versa and by limited data on the duration of anti-CHIKV immunoglobulin M positivity following acute infection. Molecular diagnostics can be used to detect and differentiate ZIKV, CHIKV, and DENV in the acute phase, and real-time reverse-transcription polymerase chain reaction (rRT-PCR) can provide quantitative data in addition to qualitative detection [6, 7, 23]. Where, these viruses cocirculate, especially in LAC, multiplex arbovirus detection including ZIKV, CHIKV, and DENV should be implemented for at-risk patients, including pregnant women, employing the same recommendations that have been issued for the screening of blood donors. A commercial multiplex molecular assay detecting ZIKV, CHIKV, and DENV has recently been accredited by the US Food and Drug Administration and may be used in routine practice [11]. In addition to these arboviruses, as mentioned, YFV, MAY, OROV as well different viral encephalitis viruses, should be also considered in this emerging scenario where is difficult to predict which of them will be the main arbovirus causing epidemics [24–26]. Given this complex scenario, any in public health and infectious diseases would be "paranoid" and trying to know who is the next one arbovirus that is knocking at our house door, wishing to be protected from its arrival, asking ourselves, "who can it be now?," as the title and lyrics of Collin Hay, in the 1981 pop song recorded by the Australian band "Men at Work." In our case, our fear and related anxiety is based on true facts that have been demonstrated by recent epidemics. Then and last, we need to increase our preparedness for emerging zoonotic and nonzoonotic arboviruses, their potential impacts and particularly to improve the vector control in tropical countries, as those in Latin America [24].

Till today, there are no single recommendations for management and treatment of such coinfections. Although that, general sense tends to recommend to orient the management to the control of the more severe of the infecting pathogens, e.g., DENV, but certainly not only acute, but also chronic implications of such coinfections (such as ZIKV and CHIKV), should be considered and followed up properly on time in order to mitigate on time their consequences [13, 27–36].

3. Cocirculation and coinfection of ZIKV with nonarboviral pathogens

Arbovirus coinfections are not the only coinfections reported. In some patients with other pathologies, an incidental diagnosis of ZIKV has been made. This situation is observed mainly in some tropical regions, where the requests for laboratory diagnosis of any arboviral infection (ZIKV, CHIKV, and DENV) in a hospitalized patient are systematically tested, as part of the public health surveillance scheme. Cocirculation of pathogens that cause acute febrile illness (AFI) complicates clinical diagnosis of patients, which can result in delays in initiating lifesaving medical interventions, as a fatal case of leptospirosis and ZIKV coinfection reported in Puerto Rico, in which ZIKV infection masked leptospirosis [37–39]. About malaria coinfection, it is described that a malaria infection may interfere with the specificity diagnosis laboratory for ZIKV, in that way, malaria may have led to false ZIKV-ELISA positives. However, the possibility of coinfection with malaria in endemic areas should be considered, tested, and treated [40, 41].

The broad symptomatic spectrum of some chronic diseases can mask the presences of other diseases, including infections. Therefore, it is important to describe the potential role played by sexually transmitted infections, mainly acute human immunodeficiency (HIV) infections, which usually debuts like fever and not specific rash which can be caused by a multitude of other pathogens, such as ZIKV, which can even be transmitted by sexual activity [42–45].

With the human immunodeficiency virus (HIV), there have been concerns about its implications. As HIV-infected adults with severe immunosuppression (e.g., a low CD4 cell count or an AIDS defining illness) experience more severe complications with infections in general, close clinical monitoring of Zika virus infection should be considered in these situations. More research is needed. Until now, there are only few case reports in the literature, but these have occurred mainly in patients under antiretroviral therapy (ARV), then with good levels of CD4 [43, 44, 46–52].

Finally, a very recent study in Colombia confirmed that coinfections would occur more than expected [53], among 157 patients in the Colombian-Venezuelan border, they found 7.6% with dengue and chikungunya, 6.4% with dengue and ZIKV, 5.1% with chikungunya and ZIKV and 1.9% with dengue, chikungunya, and Zika.

4. Conclusions

Diagnostic and treatment guidelines for those patients with simultaneous viral infections need to be urgently developed and tested in order to avoid delays in the diagnosis and associated mortality Rico [39]. The cocirculation of CHIKV, DENV, and ZIKV among existing ecological niches in LAC is a major public health challenge that requires efforts in understanding the transmission dynamics, the spectrum of clinical manifestations, health outcomes, and long-term sequelae of those coinfected with any of these emerging arboviruses [1, 11, 24]. The incidence and/or prevalence of coinfections is still a matter of concern for clinical reasons, but, its epidemiological assessment has not been furtherly studied even in the context of recent epidemics. Efforts should focus on the necessity to contain the ongoing concurrent and future epidemics and to maintain strict and continued surveillance programs to monitor the spread of these viruses as well as the introduction of newly emergent pathogens [1, 8, 11]. In the field as well as in low-income and remote areas, clinicians should take into consideration the overlapping clinical features shared among these agents as well as the possibility of coinfection in their differential diagnosis. Hopefully, clinical tools, such as the use of the term "ChikDenMaZika syndrome" [1], will provide clinicians with a useful mnemonic tool that would aid in narrowing-down diagnosis when faced with arboviral-like disease symptoms such as fever, maculopapular rash, arthralgias, myalgias, and nonpurulent conjunctivitis (or conjunctival hyperemia). Such multiagent targeted approach in clinical diagnostics should also be extrapolated to the laboratory bench by improving the usage of multiplex RT-PCR diagnostic platforms for arboviruses in returning travelers, as well as residents of endemic areas, given the increasing reported frequency of cocirculation of multiple arboviruses and its emerging threat in tropical regions.

Author details

Jorge A. Sánchez-Duque[1], Alfonso J. Rodríguez-Morales[1,2,3,4,5,6,7]*, Adriana M. Trujillo[1], Jaime A. Cardona-Ospina[1,2,6] and Wilmer E. Villamil-Gómez[3,6,8,9]

*Address all correspondence to: arodriguezm@utp.edu.co

1 Public Health and Infection Research Group, Faculty of Health Sciences, Universidad Tecnologica de Pereira, Pereira, Risaralda, Colombia

2 Infection and Immunity Research Group, Faculty of Health Sciences, Universidad Tecnologica de Pereira, Pereira, Risaralda, Colombia

3 Committee on Zoonoses, Tropical Medicine and Travel Medicine, Asociación Colombiana de Infectología, Bogotá, DC, Colombia

4 Working Group on Zoonoses, International Society for Chemotherapy, Aberdeen, United Kingdom

5 Committee on Travel Medicine, Pan-American Association of Infectious Diseases, Panama City, Panama

6 Colombian Collaborative Network for Research on Zika and Other Arboviruses (RECOLZIKA), Pereira, Risaralda, Colombia

7 Universidad Privada Franz Tamayo (UniFranz), Cochabamba, Bolivia

8 Infectious Diseases and Infection Control Research Group, Hospital Universitario de Sincelejo, Sincelejo, Sucre, Colombia

9 Programa del Doctorado de Medicina Tropical, SUE Caribe, Universidad del Atlántico, Barranquilla, Colombia

References

[1] Paniz-Mondolfi AE, Rodriguez-Morales AJ, Blohm G, Marquez M, Villamil-Gomez WE. ChikDenMaZika syndrome: The challenge of diagnosing arboviral infections in the midst of concurrent epidemics. Annals of Clinical Microbiology and Antimicrobials. 2016;**15**(1):42

[2] Rodriguez-Morales AJ, Espinoza-Flores LA. Should we be worried about sexual transmission of Zika and other arboviruses? International Maritime Health. 2017;**68**(1):68-69

[3] Rodríguez-Morales AJ, Sánchez-Duque JA, Anaya J-M. Respuesta a: Alfavirus tropicales artritogénicos. Reumatología Clínica. 2017 [Epub ahead of print]

[4] Brasil P, Calvet GA, Siqueira AM, Wakimoto M, de Sequeira PC, Nobre A, et al. Zika virus outbreak in Rio de Janeiro, Brazil: Clinical characterization, epidemiological and virological aspects. PLoS Neglected Tropical Diseases. 2016;**10**(4):e0004636

[5] Badolato-Corrêa J, Sánchez-Arcila JC, de Souza A, Manuele T, Santos Barbosa L, Conrado Guerra Nunes P, et al. Human T cell responses to Dengue and Zika virus infection compared to Dengue/Zika coinfection. Immunity, Inflammation and Disease. 2017 [Epub ahead of print]

[6] Pessôa R, Patriota JV, de Souza ML, Felix AC, Mamede N, Sanabani SS. Investigation into an outbreak of dengue-like illness in Pernambuco, Brazil, revealed a cocirculation of Zika, chikungunya, and dengue virus type 1. Medicine. 2016;**95**(12):e3201

[7] Waggoner JJ, Gresh L, Vargas MJ, Ballesteros G, Tellez Y, Soda KJ, et al. Viremia and clinical presentation in Nicaraguan patients infected with Zika virus, chikungunya virus, and dengue virus. Clinical Infectious Diseases. 2016;**63**(12):1584-1590

[8] Rückert C, Weger-Lucarelli J, Garcia-Luna SM, Young MC, Byas AD, Murrieta RA, et al. Impact of simultaneous exposure to arboviruses on infection and transmission by *Aedes aegypti* mosquitoes. Nature Communications. 2017;**8**:pii15412

[9] Dupont-Rouzeyrol M, O'Connor O, Calvez E, Daures M, John M, Grangeon J-P, et al. Co-infection with Zika and dengue viruses in 2 patients, New Caledonia, 2014. Emerging Infectious Diseases. 2015;**21**(2):381

[10] Villamil-Gómez WE, González-Camargo O, Rodriguez-Ayubi J, Zapata-Serpa D, Rodriguez-Morales AJ. Dengue, chikungunya and Zika co-infection in a patient from Colombia. Journal of Infection and Public Health. 2016;**9**(5):684-686

[11] Rodriguez-Morales AJ, Villamil-Gómez WE, Franco-Paredes C. The arboviral burden of disease caused by co-circulation and co-infection of dengue, chikungunya and Zika in the Americas. Travel Medicine and Infectious Disease. 2016;**14**(3):177-179

[12] Iovine NM, Lednicky J, Cherabuddi K, Crooke H, White SK, Loeb JC, et al. Coinfection with Zika and dengue-2 viruses in a traveler returning from Haiti, 2016: Clinical presentation and genetic analysis. Clinical Infectious Diseases. 2016:ciw667

[13] Consuegra-Rodriguez MP, Hidalgo-Zambrano DM, Vasquez-Serna H, Jimenez-Canizales CE, Parra-Valencia E, Rodriguez-Morales AJ. Post-chikungunya chronic inflammatory rheumatism: Follow-up of cases after 1 year of infection in Tolima, Colombia. Travel Medicine and Infectious Disease. 2018;**21**:62-68

[14] Villamil-Gómez WE, Rodríguez-Morales AJ, Uribe-García AM, González-Arismendy E, Castellanos JE, Calvo EP, et al. Zika, dengue, and chikungunya co-infection in a pregnant woman from Colombia. International Journal of Infectious Diseases. 2016;**51**:135-138

[15] Brito CA, Azevedo F, Cordeiro MT, Marques ET Jr, Franca RF. Central and peripheral nervous system involvement caused by Zika and chikungunya coinfection. PLoS Neglected Tropical Diseases. 2017;**11**(7):e0005583

[16] Cardoso CW, Paploski IA, Kikuti M, Rodrigues MS, Silva MM, Campos GS, et al. Outbreak of exanthematous illness associated with Zika, chikungunya, and dengue viruses, Salvador, Brazil. Emerging Infectious Diseases. 2015;**21**(12):2274

[17] Castellanos JE, Neissa JI, Camacho SJ. Dengue virus induces apoptosis in SH-SY5Y human neuroblastoma cells. Biomedica: Revista del Instituto Nacional de Salud. 2016;**36**:156-158

[18] Villamil-Gomez WE, Sanchez-Herrera AR, Hernandez H, Hernandez-Iriarte J, Diaz-Ricardo K, Castellanos J, et al. Guillain-Barre syndrome during the Zika virus outbreak in Sucre, Colombia, 2016. Travel Medicine and Infectious Disease. 2018;**16**:62-63

[19] Tang X, Zhao S, Chiu APY, Wang X, Yang L, He D. Analysing increasing trends of Guillain-Barre syndrome (GBS) and dengue cases in Hong Kong using meteorological data. PLoS One. 2017;**12**(12):e0187830

[20] Balavoine S, Pircher M, Hoen B, Herrmann-Storck C, Najioullah F, Madeux B, et al. Guillain-Barre syndrome and chikungunya: Description of all cases diagnosed during the 2014 outbreak in the French West Indies. The American Journal of Tropical Medicine and Hygiene. 2017;**97**(2):356-360

[21] Villamil-Gomez W, Silvera LA, Paez-Castellanos J, Rodriguez-Morales AJ. Guillain-Barre syndrome after chikungunya infection: A case in Colombia. Enfermedades Infecciosas y Microbiología Clínica. 2016;**34**(2):140-141

[22] Wielanek AC, Monredon JD, Amrani ME, Roger JC, Serveaux JP. Guillain-Barre syndrome complicating a chikungunya virus infection. Neurology. 2007;**69**(22):2105-2107

[23] Balmaseda A, Stettler K, Medialdea-Carrera R, Collado D, Jin X, Zambrana JV, et al. Antibody-based assay discriminates Zika virus infection from other flaviviruses. Proceedings of the National Academy of Sciences. 2017;**114**(31):8384-8389

[24] Rodríguez-Morales AJ, Sánchez-Duque JA. Preparing for next arboviral epidemics in Latin America; who can it be now?—Mayaro, Oropouche, West Nile or Venezuelan equine encephalitis viruses. Journal of Preventive Epidemiology. 2018;**3**(1):e01

[25] Patino-Barbosa AM, Bedoya-Arias JE, Cardona-Ospina JA, Rodriguez-Morales AJ. Bibliometric assessment of the scientific production of literature regarding Mayaro. Journal of Infection and Public Health. 2016;**9**(4):532-534

[26] Rodriguez-Morales AJ, Paniz-Mondolfi AE, Villamil-Gomez WE, Navarro JC. Mayaro, Oropouche and Venezuelan equine encephalitis viruses: Following in the footsteps of Zika? Travel Medicine and Infectious Disease. 2017;**15**:72-73

[27] Alvarez MF, Bolivar-Mejia A, Rodriguez-Morales AJ, Ramirez-Vallejo E. Cardiovascular involvement and manifestations of systemic chikungunya virus infection: A systematic review. F1000Research. 2017;**6**:390

[28] Jaramillo-Martinez GA, Vasquez-Serna H, Chavarro-Ordonez R, Rojas-Gomez OF, Jimenez-Canizales CE, Rodriguez-Morales AJ. Ibague Saludable: A novel tool of information and communication technologies for surveillance, prevention and control of dengue, chikungunya, Zika and other vector-borne diseases in Colombia. Journal of Infection and Public Health. 2018;**11**(1):145-146

[29] Rodriguez-Morales AJ. Letter to the editor: Chikungunya virus infection—An update on chronic rheumatism in Latin America. Rambam Maimonides Medical Journal. 2017;**8**(1): e0013

[30] Rodriguez-Morales AJ, Carvajal A, Gerardin P. Perinatally acquired chikungunya infection: Reports from the western hemisphere. The Pediatric Infectious Disease Journal. 2017; **36**(5):534-535

[31] Rodriguez-Morales AJ, Hernandez-Moncada AM, Hoyos-Guapacha KL, Vargas-Zapata SL, Sanchez-Zapata JF, Mejia-Bernal YV, et al. Potential relationships between chikungunya and depression: Solving the puzzle with key cytokines. Cytokine. 2018;**102**:161-162

[32] Rodriguez-Morales AJ, Hoyos-Guapacha KL, Vargas-Zapata SL, Meneses-Quintero OM, Gutierrez-Segura JC. Would be IL-6 a missing link between chronic inflammatory rheumatism and depression after chikungunya infection? Rheumatology International. 2017; **37**(7):1149-1151

[33] Rodriguez-Morales AJ, Mejia-Bernal YV, Meneses-Quintero OM, Gutierrez-Segura JC. Chronic depression and post-chikungunya rheumatological diseases: Is the IL-8/CXCL8 another associated mediator? Travel Medicine and Infectious Disease. 2017;**18**:77-78

[34] Rodriguez-Morales AJ, Restrepo-Posada VM, Acevedo-Escalante N, Rodriguez-Munoz ED, Valencia-Marin M, Castrillon-Spitia JD, et al. Impaired quality of life after chikungunya virus infection: A 12-month follow-up study of its chronic inflammatory rheumatism in La Virginia, Risaralda, Colombia. Rheumatology International. 2017;**37**(10):1757-1758

[35] Villamil-Gomez WE, Rodriguez-Morales AJ. Reply: Dengue RT-PCR-positive, chikungunya IgM-positive and Zika RT-PCR-positive co-infection in a patient from Colombia. Journal of Infection and Public Health. 2017;**10**(1):133-134

[36] Zambrano LI, Sierra M, Lara B, Rodriguez-Nunez I, Medina MT, Lozada-Riascos CO, et al. Estimating and mapping the incidence of dengue and chikungunya in Honduras during 2015 using geographic information systems (GIS). Journal of Infection and Public Health. 2017;**10**(4):446-456

[37] Zambrano H, Waggoner JJ, Almeida C, Rivera L, Benjamin JQ, Pinsky BA. Zika virus and chikungunya virus coinfections: A series of three cases from a single center in Ecuador. The American Journal of Tropical Medicine and Hygiene. 2016;**95**(4):894-896

[38] Biron A, Cazorla C, Amar J, Pfannstiel A, Dupont-Rouzeyrol M, Goarant C. Zika virus infection as an unexpected finding in a leptospirosis patient. JMM Case Reports. 2016;**3**(3): e005033

[39] Neaterour P, Rivera A, Galloway RL, Negrón MG, Rivera-Garcia B, Sharp TM. Fatal *Leptospira* spp./Zika virus coinfection—Puerto Rico, 2016. The American Journal of Tropical Medicine and Hygiene. 2017;**97**(4):1085-1087

[40] Schwarz NG, Mertens E, Winter D, Maiga-Ascofaré O, Dekker D, Jansen S, et al. No serological evidence for Zika virus infection and low specificity for anti-Zika virus ELISA in malaria positive individuals among pregnant women from Madagascar in 2010. PLoS One. 2017;**12**(5):e0176708

[41] Benelli G, Mehlhorn H. Declining malaria, rising of dengue and Zika virus: Insights for mosquito vector control. Parasitology Research. 2016;**115**(5):1747-1754

[42] Patiño-Barbosa AM, Medina I, Gil-Restrepo AF, Rodriguez-Morales AJ. Zika: Another sexually transmitted infection? Sexually Transmitted Infections. 2015;**91**(5):359

[43] Penot P, Brichler S, Guilleminot J, Lascoux-Combe C, Taulera O, Gordien E, et al. Infectious Zika virus in vaginal secretions from an HIV-infected woman, France, August 2016. Eurosurveillance. 2017;**22**(3):pii3044

[44] Hirschel T, Steffen H, Pecoul V, Calmy A. Blinded by Zika? A missed HIV diagnosis that resulted in optic neuropathy and blindness: A case report. BMC Research Notes. 2017; **10**(1):664

[45] Rabelo K, Fernandes RCDSC, Souza L, Louvain De Souza T, Santos F, Nunes P, et al. Placental histopathology and clinical presentation of severe congenital Zika syndrome in an HIV-exposed uninfected infant. Frontiers in Immunology. 2017;**8**:1704

[46] Barreiro P. First case of Zika virus infection in a HIV+ patient. AIDS Reviews. 2016;**18**(2):112

[47] Barreiro P, Evolving RNA. Virus pandemics: HIV, HCV, Ebola, Dengue, chikungunya, and now Zika! AIDS Reviews. 2016;**18**(1):54-55

[48] Calvet GA, Filippis AM, Mendonca MC, Sequeira PC, Siqueira AM, Veloso VG, et al. First detection of autochthonous Zika virus transmission in a HIV-infected patient in Rio de Janeiro, Brazil. Journal of Clinical Virology: The Official Publication of the Pan American Society for Clinical Virology. 2016;**74**:1-3

[49] Joao EC, Gouvea MI, Teixeira ML, Mendes-Silva W, Esteves JS, Santos EM, et al. Zika virus infection associated with congenital birth defects in a HIV-infected pregnant woman. The Pediatric Infectious Disease Journal. 2017;**36**(5):500-501

[50] Tsuyuki K, Gipson JD, Barbosa RM, Urada LA, Morisky DE. Preventing syndemic Zika virus, HIV/STIs and unintended pregnancy: Dual method use and consistent condom use among Brazilian women in marital and civil unions. Culture, Health & Sexuality. 2017:1-17. Epub ahead of print

[51] Wainberg MA, Lever AM. HIV and Zika: When will we be able to end these epidemics? Retrovirology. 2016;**13**(1):80

[52] Zorrilla CD, Mosquera AM, Rabionet S, Rivera-Vinas J. HIV and ZIKA in pregnancy: Parallel stories and new challenges. Obstetrics and Gynecology International. 2016;**5**(6): pii00180

[53] Carrillo-Hernandez MY, Ruiz-Saenz J, Villamizar LJ, Gomez-Rangel SY, Martinez-Gutierrez M. Co-circulation and simultaneous co-infection of dengue, chikungunya, and Zika viruses in patients with febrile syndrome at the Colombian-Venezuelan border. BMC Infectious Diseases. 2018;**18**:61